Mosses of Singapore and Malaysia

ANNE JOHNSON

SINGAPORE UNIVERSITY PRESS

Printed by Welco Printers Pte. Ltd.
Tel: 2833587

Preface

Southeast Asia has a remarkable heritage of aesthetically attractive plants and animals. The larger representatives of these have already been the subject of a number of popular books. Most nature enthusiasts and students have, however, overlooked small representatives such as mosses whose structure is so unusual and interesting. Moss study presents two kinds of problems which have, I believe, led to their neglect: (1) they are at the limit of normal unaided vision so that their structure can scarcely be appreciated fully with the naked eye—a properly used 10× or 15× hand-lens is essential, and the use of a low-powered compound microscope is desirable; (2) mosses have no commonly accepted popular names. This is a stumbling-block to non-biologists.

The mosses of Singapore and Malaysia form an attractive and fascinating group of plants, but there are no books available in English for the beginner. This book includes a readily available account of a selection of these tiny, beautifully-structured plants. English names based on the original scientific names, where possible, have been provided. Literal meanings of the scientific names are given in the Appendix. About 500 species of mosses found in Singapore and Malaysia have been recorded. Many of these are little known and have only been collected a few times. This elementary account includes all recorded Singapore species, as well as some others found only in Malaysia. Singapore species are indicated by an asterisk. Except for limestone districts, the mosses of the lowlands of Malaysia are similar to those of Singapore.

Contents

Introduction 1
How to study mosses 4
Economic uses of mosses 7
List of mosses 8
Key to moss genera 13
Bog moss family (Sphagnaceae) 23
Fork and swan moss family (Dicranaceae) 24
White moss family (Leucobryaceae) 39
Pouched hood moss family (Calymperaceae) 49
Horizontal tooth moss family (Syrrhopodontaceae) 55
Split tooth moss family (Fissidentaceae) 64
Pott moss family (Pottiaceae) 71
Bryum family (Bryaceae) 77
Brush moss family (Rhizogoniaceae) 77
Weber's moss family (Weberaceae) 78
Hair mosses (Polytrichaceae) 79
Straight hair mosses (Orthotrichaceae) 82
Necker's moss family (Neckeraceae) 83
Hooker's moss family (Hookeriaceae) 88
Conifer moss family (Thuidiaceae) 94
Hypnon family (Hypnaceae) 97
Signal leaf family (Sematophyllaceae) 107
Ruined tooth moss family (Rhegmatodontaceae) 113
Glossary 114
Appendix: Hints on the meanings of names 119
References 124
Index 125

INTRODUCTION

Mosses belong to the group of plants which never have flowers or seeds (*flowerless plants*). This group also includes seaweeds, mushrooms, lichens, liverworts, and ferns. Most mosses are small and insignificant, but some may be as much as 35 cm tall. In some parts of the world clumps of moss may become so large that they are mistaken for sheep. Moss plants do not have veins to transport water and food from one part of the plant body to another. Instead all parts of the plant can absorb water. Most mosses can dry up and yet still remain alive for many months. When wetted they will begin to grow again. Mosses have been known to remain alive in a dry state for as long as five years in a herbarium.

Mosses never have roots. Instead the plant bears branched threads called *rhizoids*. While these rhizoids can absorb water, their main function is anchorage since, as explained above, all cells of mosses are absorptive. Mosses are found most abundantly where there is a sufficient supply of water. Some species require shady conditions, while others thrive in full sun. In Singapore and Malaysia, they are more abundant in the forest than in open areas. In Malaysia extensive mossy areas are found in the mountains.

The major groups of mosses are based on the characters of the *capsule* which contains the spores, but this is frequently absent especially in the lowlands. Mosses are also distinguished by features of their stem and leaves. If these are carefully observed it will be seen that they vary tremendously in shape, size, arrangement, colour, etc. Since mosses are at the limit of visibility, the use of a hand-lens is essential. This must be held close to one eye, while the other eye remains open. Although this takes a bit of practice, it is well worth persisting. The plant is then brought up towards the eye until it is in focus.

Since mosses are sometimes restricted to certain habitats, the place where a moss was found should be carefully noted by the collector. Some mosses never grow on the bark of trees, while others always do; some grow on rocks or the side of drains, etc. Each specimen collected should be carefully labelled to indicate the place it was collected and the date.

Mosses are green plants since their leaves usually have a large number of chloroplast-containing cells. Although the word "leaf" is used, it must be realized that the leaf of a moss is quite different from that of a higher plant like a mango or a balsam. Leaves of higher plants are thicker structures with a leaf-blade (*lamina*) many cells thick. Higher plant leaves have veins which can transport water and other substances.

Moss leaves have a green lamina which is only one cell thick. If you examine the lamina under the microscope you will not see any overlapping of the cell walls, since there is only one layer. A moss leaf does not have any veins to transport water. It may have a mid-rib (nerve) but this is simply made of several layers of thickened or elongated cells for support and does not have any function in transporting water or other substances. A few mosses have two nerves and quite a large number have none at all. The nerve of a moss is defined as any central part of the leaf which is more than one cell thick. In the White Moss family (Leucobryaceae) most of the leaf is made up of several layers of dead cells (*leucocysts*) in which one or more layers of green cells (*chlorocysts*) are hidden. This means that in this family most of the leaf is considered to be the nerve.

While mosses are essentially land plants, they depend on a film of water for the transport of the male cell (*spermatozoid*) to the female cell. Apart from Bog Moss (*Sphagnum*), only three genera of mosses occur in fresh water and none of these have been reported from Singapore or Malaysia. Most mosses cannot tolerate salt water or sea spray since this will damage the delicate lamina cells on the leaf. Only one species, *Fissidens crassinervis,* has been reported from mud lobster burrows in mangrove. Apart from this limitation of abhorrence of salt water, mosses are hardy plants growing in a great variety of places as judged by light or humidity conditions, including caves, bright sunlight, and damp tropical forest. The substratum on which they grow varies from ordinary soil, to detritus, rubble, bark, or rocks. Some species form tufts of just a few individuals which may even become separated so the individual plants are solitary. Most mosses are, however, gregarious with a large number occurring together to form cushions, tufts, carpets, or mats. Both upright and creeping mosses may branch, either sparsely or profusely. Branches can be quite irregular and very untidy, or regular forming a feather-like (*pinnate*) pattern. Some of the epiphytic mosses may hang down from the branches of trees.

Mosses are divided into two groups according to the type of branching and the position of the capsule in relation to the apex.

GROUP I ACROCARPOUS MOSSES

Acrocarpous mosses have upright stems which are either unbranched (*simple*) or sparsely branched. The sex organs are borne terminally on the main stem, and their formation prevents further growth from the apex. If the plant is perennial, new growth in the next season is by a branch which arises laterally below the apex. This is called branching by *innovation.* Acrocarpous mosses are found as small groups of individual plants or as cushions or tufts. In a cushion or tuft the dead older parts of the upright plants are found at the base, forming a mass of dead matter which accumulates from season to season. Capsules are borne at the apex of female mosses after fertilization of the female sex organ by a spermatozoid.

GROUP 2 PLEUROCARPOUS MOSSES

Pleurocarpous mosses have stems which creep along the surface of the ground, rock, or bark of tree, or hang down from a tree. Branches are usually numerous. They may be creeping like the main stem; erect (at right angles to the main stem); ascending (growing away from the substratum at an angle less than 90°); or hanging down (*pendent*). These mosses produce their sex organs on special short branches which are produced laterally from the main stem. Because of this the main stem can continue its growth indefinitely, and production of capsules by the female plant does not stop the main stem growth. Most of these mosses form mats or carpets. A carpet is larger and more extensive than a mat. Old parts of the plants will die in the centre of a mat or carpet.

HOW TO STUDY MOSSES

The equipment needed to study mosses is not expensive. For collection you will need some small polythene bags, a pen-knife, some pieces of white paper for labels, a pencil, a notebook, and a hand-lens. The lens may be suspended on a chain or a piece of string round your neck so that both your hands are free if you are doing extensive climbing or need your hands for collection. Each specimen should, if possible, consist of only one species taken separately and placed in its own polythene bag, labelled, and sealed by tying up the bag. The label, written in pencil, should note the number and habitat of the specimen, together with the date. If you have never numbered your collection before, start with number 1 and continue the series throughout your life. Never use the same number twice. It is better to continue the series to include both mosses and other kinds of plants (if you are also collecting these), although some collectors form a distinct series of moss numbers. Such a series could be numbered M1, M2, M3, etc. Difficulties may arise if you later find that something you have collected is not a moss. This is one reason why it is better to have a single series of numbers.

Mosses collected in polythene bags will keep for several days if you are on a long collecting trip. Since mosses do not have roots, they are not restricted in travelling from one country to another. Most customs officials will overlook them. If the mosses are very wet when collected the paper label may decay after a week or so. Some collectors prefer to use white cardboard labels to avoid this. Make sure your polythene bags are properly sealed. Do not allow the mosses to dry in the bags without pressing. After returning home from your collecting trip, mosses should be dried between sheets of newspaper in a plant-press or under a heavy piece of furniture. Newspapers should be changed every day until the mosses are completely dry. Some collectors complete the pressing by placing the mosses under a carpet over which people walk.

Dried specimens should be completely flat. They should be placed in ungummed envelopes or folded pieces of paper, and stored in boxes. Large mosses may be sewed on to large sheets of herbarium paper and be kept with higher plants. All specimens should be clearly labelled with the name of the plant (if known), collector's name, collector's number, place collected, etc. Unlike higher plants there is no need to poison moss specimens since they are not attacked by fungi if completely dry. They should, however, be kept with a few napthalene balls in an insect-proof cupboard. You should avoid using staples, paper-clips, or sellotape in preparing moss packets since they soon deteriorate in the tropics. Specimens will keep for many years. At any time a small fragment of moss can be removed and soaked in water and examined. If you are fortunate enough to have a microscope, you can place the piece of moss on a slide with a drop of water and remove a few leaves with two needles. It is important to get complete leaves as features at the base of the leaf are often very interesting and important. It is also very difficult to gauge the real shape of a leaf if the base is missing.

Since mosses are of reasonable size, it is not essential to have a very expensive or elaborate microscrope. For ordinary examination the specimen is mounted in a drop of water on a slide, covered with a coverslip, and examined under low power. Opaque objects, like capsules, may be cut in half, mounted in plasticine with the top upwards, and examined with direct lighting under low power. Permanent slides may be made by mounting in glycerine and sealing the edge of the coverslip with sealing cement. The top half of capsules may be glued to a slide whose centre is blackened with india ink and protected by a ring mount and a coverslip. Leaves, stems, rhizoids, and protonemata may also be stained with Fast Green or other microscopic stains according to normal staining methods and mounted in euparol.

Differences between Mosses and Liverworts

Mosses are quite different from liverworts, but it is not easy to explain exactly how they differ. This is because, although mosses are a fairly uniform group, liverworts are very variable and some are quite similar to mosses. Since there are very few characteristics common to all liverworts, comparisons are difficult. We can however use the following guidelines:

1. All mosses have both stem and leaves; therefore if you have a flat thallus-like plant, it cannot be a moss.

2. Mosses have their leaves arranged normally in two, three, four, eight, or numerous rows; or they are spirally arranged. Leafy liverworts normally have two rows of leaves: one at each side, plus an additional row of underleaves (*amphigastria*). In these leafy liverworts, the lateral leaves are placed at an angle (*obliquely inserted*) so that they usually overlap; while the underleaves are placed across the stem (*transversely inserted*) so that their base is at right angles to the main axis.

3. Moss leaves are never lobed or divided into two parts, while some leafy liverworts are so divided.

4. Moss leaves may have a mid-rib (*nerve*) or even two nerves. Liverwort leaves never have a nerve, although a liverwort thallus may have a mid-rib. However, not all moss leaves have nerves.

5. Moss stems often have thickened supporting cells in the centre. Leafy liverwort stems lack this support.

6. In a moss the outer part of the female organ (*archegonium*) breaks into two parts as the seta elongates. One part is carried up to form the *calyptra* while the other remains at the base of the seta to form the *vaginula*. In liverworts the seta does not carry up the calyptra.

7. A moss seta is a tough, long-lasting wiry stalk, often reddish or purplish in colour. Its elongation takes several days or weeks. A liverwort seta is a weak transparent structure, white in colour. Its elongation takes only a few hours, after which it collapses and dies.

8. A moss capsule is usually pear-shaped or cylindrical, green when young, and with a distinct lid (*operculum.*) which is shed on maturity to reveal the peristome teeth. These teeth usually respond to changes in humidity to open or close the mouth. A liverwort capsule is never green (except *Anthoceros*), is round or globe shaped, and often splits into four parts when ripe. It does not usually have a lid.

9. Spores of both mosses and liverworts germinate to give *protonemata,* but that of a moss is much more extensive and gives rise to numerous moss plants. A liverwort protonema gives rise to a single plant.

10. If you have a microscope, there is an infallible way of telling if you have a moss or a liverwort. Look at the rhizoids. If they are many-celled and branched, the specimen is a moss. If the rhizoids consist of one cell which is unbranched, you have liverwort.

To Decide Whether Your Specimen is a Moss or a Liverwort

1. If it does not have any leaves, but is just a flat piece of tissue (thallus)	*It must be a liverwort.* There are no thallose mosses.
2. If it is leafy	If the leaves are spirally arranged or consist of 2–8 rows on similar leaves; and the capsule (if present) has a lid with peristome teeth below it, with a thick coloured seta, a calyptra over the lid, and rhizoids of many cells, *It is a moss.*
	If the leaves are in two lateral rows which are obliquely inserted, plus one row of transversely inserted underleaves; if it creeps on the ground or the bark of trees and has a capsule (if present) without a lid, without peristome teeth, with a calyptra not carried up; and rhizoids each of one very long cell, *It is a leafy liverwort.*

ECONOMIC USES OF MOSSES

Mosses are not plants of great direct economic importance but, like all other living things, play an important role in the ecological balance of the environment. Some mosses have, however, been of direct service to man, for example, Bog Moss (*Sphagnum*). Locally this moss is only found at the tops of high mountains and in some peat swamp forests, but it is much more widespread in bogs in temperate regions. Its accumulation there gives rise to peat which is used as a fuel, for example, in Ireland. The dried *Sphagnum* plant has a remarkable power of absorbing water, due to the large number of empty cells of which the plant body is composed. Because of this property, *Sphagnum* has been used in certain industries as an absorptive agent, and in horticulture as a medium for the germination of small seedlings. During the First World War (1914–18) *Sphagnum* was used in the battlefields in France as an emergency dressing for wounds. Owing to the fact that this Bog Moss grows only under extremely acid conditions, it was found to partially sterilize wounds and to act as a styptic for bleeding.

In some parts of the world, the Hair Moss (*Polytrichum*) has been used to make brushes, little brooms, lamp-wicks, and attractive woven baskets. *Polytrichum* is closely related to the Malaysian *Pogonatum.*

Burkhill (1966) has reported that *Calymperes delessertii* and other mosses were used by the Malays for mattress stuffing. On the east coast of Peninisular Malaysia, *Leucophanes octoblephanioides* and other members to the White Moss family (Leucobryaceae) are used to stuff cushions. Expanses of moss more than one metre in diameter are often found.

Moss gardens cultivated with miniature trees are found in Japan.

LIST OF MOSSES INCLUDED IN THIS BOOK

Mosses found in Singapore are indicated with an asterisk.

BOG MOSS FAMILY
(Sphagnaceae)

Flaccid-Leaved Bog Moss	*Sphagnum flaccidifolium*
Junghuhn's Bog Moss	*Sphagnum junghuhnianum*

FORK AND SWAN MOSS FAMILY
(Dicranaceae)

Wilson's Fork Moss	*Wilsoniella pellucida*
Vase Fork Moss	*Garckea phascoides*
Difficult Two-Haired Fork Moss	*Ditrichum difficile*
*Miquel's Small-Toothed Fork Moss	*Microdus miquelianus*
Compressed Fork Moss	*Dicranella coarctata*
True-Branched Swan Moss	*Campylopodium euphorocladum*
Braunfels' Fork Moss	*Braunfelsia dicranoides*
Braunfels' Nerveless Moss	*Braunfelsia enervis*
Similar Fork Moss	*Dicranoloma assimile*
Tender White-Bordered Fork Moss	*Leucoloma molle*
Hairy Swan Moss	*Campylopus comosus*
*Toothed Swan Moss	*Campylopus serratus*
Thwaites' Swan Moss	*Campylopus thwaitesii*
Richards' Swan Moss	*Campylopus richardii*
Shining Fork Moss	*Dicranodontium nitidum*

WHITE MOSS FAMILY
(Leucobryaceae)

*Holy White Moss	*Leucobryum sanctum*
*Javanese White Moss	*Leucobryum javense*
*Hooked White Moss	*Leucobryum aduncum*
*Eight-Rowed White Moss	*Leucophanes octoblepharioides*
*Pure White Moss	*Leucophanes candidum*
*Narrow-Leaved White Moss	*Leucophanes albescens*
*Jointed White Moss	*Arthrocormus schimperi*
Pointed Split-Cap White Moss	*Schistomitrium apiculatum*
Hairy Short-Branched White Moss	*Cladopodanthus pilifer*

Blume's Outer Net White Moss	*Exodictyon blumii*
Sullivant's Outer Net White Moss	*Exodictyon sullivantii*
Eight-Toothed White Moss	*Octoblepharum albidum*

POUCHED HOOD MOSS FAMILY
(Calymperaceae)

*Dozy's Pouched Hood Moss	*Calymperes dozyanum*
*Toothed Pouched Hood Moss	*Calymperes serratum*
*Long-Leaved Pouched Hood Moss	*Calymperes longifolium*
*Extended Pouched Hood Moss	*Calymperes porrectum*
*Curly Pouched Hood Moss	*Calymperes recurvifolium*
*Tahitan Pouched Hood Moss	*Calymperes tahitense*
*Hampe's Pouched Hood Moss	*Calymperes hampei*
*Nicobar Pouched Hood Moss	*Calymperes nicobarense*
*Delicate Pouched Hood Moss	*Calymperes tenerum*
*Tjibodas Pouched Hood Moss	*Calymperidium tjibodensis*

HORIZONTAL TOOTH MOSS FAMILY
(Syrrhopodontaceae)

*Rolled Horizontal Tooth Moss	*Syrrhopodon revolutus*
Gardner's Moss	*Syrrhopodon gardneri*
*Reddish Horizontal Tooth Moss	*Syrrhopodon rufescens*
*White-Sheathed Horizontal Tooth Moss	*Syrrhopodon albo-vaginatus*
*Hairy Horizontal Tooth Moss	*Syrrhopodon ciliatus*
*Spiny Horizontal Tooth Moss	*Syrrhopodon spiculosus*
*Yellow Horizontal Tooth Moss	*Syrrhopodon croceus*
*Mueller's Horizontal Tooth Moss	*Syrrhopodon muelleri*
*Ridley's Horizontal Tooth Moss	*Syrrhopodon ridleyi*
*Dense Horizontal Tooth Moss	*Syrrhopodon confertus*
*Yellow Gate Moss	*Thyridium flavum*
*Bundled Gate Moss	*Thyridium fasciculatum*
*Cardot's Gate Moss	*Thyridium cardotii*
*Creeping Gate Moss	*Thyridium repens*
*Man's Gate Moss	*Thyridium manii*
*Wavy Gate Moss	*Thyridium undulatum*

SPLIT TOOTH MOSS FAMILY
(Fissidentaceae)

*Ceylonese Split Tooth Moss	*Fissidens ceylonensis*
*Woodland Split Tooth Moss	*Fissidens sylvaticus*
*Thick-Nerved Split Tooth Moss	*Fissidens crassinervis*
*Mitten's Split Tooth Moss	*Fissidens mittenii*

POTT MOSS FAMILY
(Pottiaceae)

Snow-Bordered Pott Moss	*Chionoloma latifolium*
Veiled-Mouth Pott Moss	*Hymenostomum edentulum*
Javanese Pott Moss	*Hyophila javanica*
Inrolled Pott Moss	*Hyophila involuta*
*Eastern Pott Moss	*Semibarbula orientalis*
Louisiaden Pott Moss	*Barbula louisiadum*
*Curved-Beard Pott Moss	*Hydrogonium arcuatum*
Sarawak Pott Moss	*Trichostomum sarawakensis*

BRYUM FAMILY
(Bryaceae)

*Crown Moss	*Bryum coronatum*

BRUSH MOSS FAMILY
(Rhizogoniaceae)

*Spiny Brush Moss	*Rhizogonium spiniforme*
*Broad-Leaved Brush Moss	*Rhizogonium latifolium*

WEBER'S MOSS FAMILY
(Weberaceae)

*Weber's Short-Tipped Moss	*Diphyscium mucronifolium*

HAIR MOSSES
(Polytrichaceae)

Large-Leaved Hair Moss	*Pogonatum macrophyllum*
Junghuhn's Hair Moss	*Pogonatum junghuhnianum*
Ruined-Tooth Hair Moss	*Racelopus pilifer*

STRAIGHT HAIR MOSSES
(Orthotrichaceae)

*Pointed Hidden-Fruit Moss	*Desmotheca apiculata*
*Curved-Leaf Straight Hair Moss	*Macromitrium incurvifolium*

NECKER'S MOSS FAMILY
(Neckeraceae)

Elegant Inner-Hair Moss	*Endotrichella elegans*
Necker's Joined-Toothed Moss	*Symphysodon neckeroides*
Feathery Strap-shaped Moss	*Himantocladium plumula*
Elongate Aerial Moss	*Aerobryopsis longissima*
Miquel's Moss	*Meteorium miquelianum*
Profusely Fertile Moss	*Floribundaria floribunda*
*Small Mucronate Feather Moss	*Pinnatella mucronata*

HOOKER'S MOSS FAMILY
(Hookeriaceae)

Yellow Bent-Tooth Moss	*Campylodontium flavescens*
Pointed Two-Row Moss	*Distichophyllum cuspidatum*
Spoon-shaped Two-Row Moss	*Distichophyllum spathulatum*
*Slender-Stem Two-Row Moss	*Distichophyllum gracilicaule*
*Schmidt's Two-Row Moss	*Distichophyllum schmidtii*
Slender-Fruit Hairy-Cap Moss	*Chaetomitrium leptopoma*
Erect-Beak Hairy-Cap Moss	*Chaetomitrium orthorrhynchum*
Elongate Hairy-Cap Moss	*Chaetomitrium elongatum*
*Beautiful-Veined Moss	*Callicostella prabaktiana*

CONIFER MOSS FAMILY
(Thuidiaceae)

*Covered Axe Moss	*Pelekium velatum*
*Two-Row Axe Moss	*Pelekium bifarium*

HYPNON FAMILY
(Hypnaceae)

*Necker's Oblique-Fruit Moss	*Plagiothecium neckeroideum*
*Buitenzorg Folded-Fruit Moss	*Ectropothecium buitenzorgii*
*Moritz's Folded-Fruit Moss	*Ectropothecium moritzii*
*Singapore Folded-Fruit Moss	*Ectropothecium singapurensis*

*Obliquely-Inserted Folded-Fruit Moss	*Ectropothecium incubans*
*Monument Folded-Fruit Moss	*Ectropothecium monumentorum*
*Minute Branched Symmetrical Moss	*Isopterygium minutirameum*
*White Symmetrical Moss	*Isopterygium albescens*
*Banka Symmetrical Moss	*Isopterygium bancanum*
*Miquel's Bladder Moss	*Vesicularia miquelii*
*Net Bladder Moss	*Vesicularia reticulata*
*Dubious Bladder Moss	*Vesicularia dubyana*
*Kurz's Bladder Moss	*Vesicularia kurzii*
*Spiny Beaked Moss	*Acanthorrhynchium papillatum*
*Uncovered Nipple Moss	*Taxithelium instratum*
*Gottsch's Nipple Moss	*Taxithelium gottscheanum*
*Keri's Nipple Moss	*Taxithelium kerianum*
*Equal Branch Nipple Moss	*Taxithelium isocladum*
*Hairy Nipple Moss	*Taxithelium capillipes*

SIGNAL LEAF FAMILY
(Sematophyllaceae)

Reddish Terminal Pore Moss	*Acroporium rufum*
*Rough-Leaved Terminal Pore	*Acroporium asperifolium*
*Triple Large Moss	*Trismegistia lancifolia*
*Rigid Triple Large Moss	*Trismegistia rigida*
*Small Fruit Moss	*Meiothecium microcarpum*
*Jagor's Small Fruit Moss	*Meiothecium jagorii*
*Warburg's Narrow Fruit Moss	*Warburgiella leptocarpum*
*Rotten Wood Signal Moss	*Sematophyllum saproxylophilum*
*Short-Branched Signal Moss	*Sematophyllum microcladiellum*
*Luxuriant Rhaphe Moss	*Rhaphidostichum luxurians*
*Bruch's Rhaphe Moss	*Rhaphidostichum bruchii*
*Bosch's Column Moss	*Trichosteleum boschii*
*Singapore Column Moss	*Trichosteleum singapurense*
*One-Row Column Moss	*Trichosteleum monostichum*

RUINED TOOTH MOSS FAMILY
(Rhegmatodontaceae)

Mueller's Large Membrane Moss	*Macrohymenium muelleri*

KEY TO MOSS GENERA

Like other plants and animals, mosses have names consisting of two words: a generic name followed by a specific name. There may be several species belonging to one genus, for example, White Moss, *Leucobryum*, is a genus with several species in this region, including, Holy White Moss, *Leucobryum sanctum*; Javanese White Moss, *Leucobryum javense*; and Hooked White Moss, *Leucobryum aduncum*.

The following key presents alternatives by which the genera described in this book may be identified. Each pair of alternatives has the same number given on the left-hand side of the page. Start at the two alternatives numbered 1. Choose the one which is more appropriate for your specimen. If you choose the first alternative, your specimen belongs to the genus *Fissidens*. You can find details of species in this genus on pp. 66–71. If you choose the second alternative under 1, it leads to the number 2 on the right-hand side. Follow the same number on the left-hand side and choose between the pair of alternatives, as before. Continue in this manner until you have found the name of the genus. Check the features of your moss with the description in the main body of the text. Note also the localities and whether they correspond to the kind of place you found it. Since so few people have collected in this part of the world it is very likely you will establish new records. However, in general, you will not normally expect to find mountain mosses in the lowlands or limestone mosses outside limestone areas. Mosses found in Singapore are indicated with an asterisk. Most of these mosses are also found in the lowlands of Malaysia. There is a simplified key for mosses found in Singapore and the lowlands of Malaysia on pages 19–22.

1 Leaves in two alternate rows .... FISSIDENS* (p. 67)
1 Leaves in three rows or spirally arranged 2

2 Peat swamp or mountain-top mosses; branches in fascicles .... SPHAGNUM (p. 24)
2 Branches not in fascicles 3

3 Plant glaucous; leaves mainly of leucocysts 4
3 Plant not glaucous; leaves without leucocysts 10

4 Thickened false mid-rib present .... LEUCOPHANES* (p. 41)
4 Without thickened false mid-rib 5

5	Leaves in three rows	ARTHROCORMUS* (p. 42)
5	Leaves in five rows or spirally arranged	6
6	Green cells exposed on leaf surfaces	EXODICTYON (p. 46)
6	Green cells not exposed	7
7	Leaves linear, thick, dorsally convex	OCTOBLEPHARUM (p. 49)
7	Leaves lanceolate or spoon-shaped, thin	8
8	Stem creeping; leaves adpressed, hair-pointed	CLADOPODANTHUS (p. 46)
8	Stem upright; leaves erect-spreading, not hair-pointed	9
9	Peristome teeth dicranoid	LEUCOBRYUM* (p. 39)
9	Peristome teeth entire	SCHISTOMITRIUM (p. 45)
10	Leaf lamina one layer of cells	11
10	Leaf lamina two layers of cells	DIPHYSCIUM* (p. 79)
11	Ventral surface of leaf with lamellae	POGONATUM (p. 79)
11	Ventral surface of leaf without lamellae	12
12	Inner basal leaf cells forming colourless sheath	13
12	Inner basal leaf cells not differentiated	16
13	Primary stem creeping	THYRIDIUM* (p. 62)
13	Primary stem erect	14
14	Upper leaves with hair-like gemmae	15
14	No gemmae present	SYRRHOPODON* (p. 56)

15 Gemmae mid-way along mid-rib .... CALYMPERIOPSIS* (p. 55)
15 Gemmae terminal on excurrent mid-rib .... CALYMPERES* (p. 50)

16 Basal leaf cells sinuose or nodulose .... CHIONOLOMA (p. 71)
16 Basal leaf cells not sinuose or nodulose 17

17 Stems erect; acrocarpous mosses 18
17 Stems creeping; pleurocarpous mosses 40

18 Upper leaf cells prosenchymatous 19
18 Upper leaf cells parenchymatous 26

19 Leaf cells rhombic or elongated rectangular 20
19 Leaf cells narrowly linear 23

20 Upper leaves forming conspicuous rosettes .... BRYUM* (p. 77)
20 Upper leaves not forming rosettes 21

21 Leaves narrowly lingulate .... WILSONIELLA (p. 27)
21 Leaves lanceolate or broadly spoon-shaped 22

22 Alar cells present .... CAMPYLOPUS* (p. 34)
22 Alar cells absent .... GARCKEA (p. 27)

23 Alar cells present 24
23 Alar cells absent 25

24 Leaves bordered .... DICRANODONTIUM (p. 38)
24 Leaves not bordered .... BRAUNFELSIA (p. 30)

25 Peristome teeth striate, divided to middle .... DICRANELLA (p. 29)
25 Peristome teeth entire, papillose .... MICRODUS* (p. 28)

26	Leaf cells wide; hexagonal or retangular	27
26	Leaf cells small; square, rounded, or hexagonal	29
27	Leaf cells elongately hexagonal	RACELOPUS (p. 80)
27	Leaf cells rectangular	28
28	Seta cygneous	CAMPYLOPODIUM (p. 30)
28	Seta not cygneous	DITRICHUM (p. 28)
29	Primary stem creeping with erect branches	30
29	Primary stem erect	31
30	Capsule on long seta	MACROMITRIUM* (p. 83)
30	Capsule immersed	DESMOTHECA* (p. 82)
31	Alar cells present	32
31	Alar cells absent	34
32	Leaves bordered	33
32	Leaves not bordered	CAMPYLOPUS* (p. 37)
33	Leaf cells papillose	LEUCOLOMA (p. 34)
33	Leaf cells smooth	DICRANOLOMA (p. 33)
34	Leaves spoon-shaped to lingulate	HYOPHILA (p. 72)
34	Leaves lanceolate to linear-subulate	35
35	Basal leaf cells rectangular or elongated	36
35	Basal leaf cells not differentiated	37
36	Peristome absent. Mouth closed by membrane	HYMENOSTOMUM (p. 72)
36	Peristome teeth present. No membrane	TRICHOSTOMUM (p. 74)
37	Leaves bordered with sharp, double teeth	RHIZOGONIUM* (p. 78)

37	Leaves not bordered, entire or tiny teeth	38
38	Mid-rib rough on dorsal side	SEMIBARBULA* (p. 73)
38	Mid-rib not rough on dorsal side	39
39	Leaves not translucent	BARBULA (p. 74)
39	Leaves translucent	HYDROGONIUM* (p. 74)
40	Alar cells absent	41
40	Alar cells present	55
41	Leaf cells parenchymatous	42
41	Leaf cells prosenchymatous	48
42	Mid-rib very prominent and double	CALLICOSTELLA* (p. 94)
42	Mid-rib single, absent, or double and inconspicuous	43
43	Leaf cells very small	44
43	Leaf cells wide hexagonal	DISTICHOPHYLLUM* (p. 89)
44	Leaf cells smooth	HIMANTOCLADIUM (p. 84)
44	Leaf cells papillose	45
45	Paraphyllia present	PELEKIUM* (p. 96)
45	Paraphyllia absent	46
46	Capsule immersed	DESMOTHECA* (p. 82)
46	Capsule with long seta	47
47	Basal cells elongated	MACROMITRIUM* (p. 83)
47	Basal leaf cells not differentiated	PINNATELLA* (p. 88)
48	Leaf cells rhombic	METEORIUM (p. 87)
48	Leaf cells usually linear	49
49	Mid-rib present	50
49	Mid-rib absent	51

50 Branch leaves transversely undulate .... AEROBRYOPSIS (p. 87)
50 Branch leaves not transversely undulate .... FLORIBUNDARIA (p. 88)

51 Leaf cells thick-walled 52
51 Leaf cells thin-walled 53

52 Capsule immersed .... ENDOTRICHELLA (p. 83)
52 Capsule on long seta .... CHAETOMITRIUM (p. 93)

53 Capsule immersed .... SYMPHYSODON (p. 84)
53 Capsule on long seta 54

54 Leaf with an indistinct border .... VESICULARIA* (p. 102)
54 Leaf without a border .... ISOPTERYGIUM* (p. 101)

55 Apical leaves rolled into a distinct point .... ACROPORIUM (p. 107)
55 Apical leaves not rolled into a distinct point 56

56 Leaf cells thick-walled 57
56 Leaf cells thin-walled 60

57 Leaf cells papillose .... ACANTHORRHYNCHIUM* (p. 103)
57 Leaf cells not papillose 58

58 Leaf cells rhombic .... MACROHYMENIUM (p. 113)
58 Leaf cells linear 59

59 Leaf margin entire .... MEIOTHECIUM* (p. 108)
59 Leaf margin dentate .... TRISMEGISTIA* (p. 107)

60 Leaf cells papillose 61
60 Leaf cells not papillose 63

61 Leaves falcate 62
61 Leaves spreading .... TAXITHELIUM* (p. 105)

62	Seta scabrous above	TRICHOSTELEUM* (p. 112)
62	Seta smooth	WARBURGIELLA* (p. 111)
63	Alar cells divided or in several rows	64
63	Alar cells not divided; in one row	66
64	Branches complanate	PLAGIOTHECIUM* (p. 97)
64	Branches not complanate	65
65	Leaves longitudinally plicate	CAMPYLODONTIUM (p. 89)
65	Leaves not longitudinally plicate	SECTROPOTHECIUM* (p. 97)
66	Apex forming an aristate point	RHAPHIDOSTICHUM (p. 112)
66	Apex blunt or with subulate point	SEMATOPHYLLUM (p. 111)

Simplified Key for Use in Singapore and the Lowlands of Malaysia

The following key may be used to identify mosses found in Singapore and the lowlands of Malaysia, except for limestone districts. It should not be used for other parts of Malaysia.

1	Leaves in two alternate rows	FISSIDENS* (p. 67)
1	Leaves in many rows or spirally arranged	2
2	Plants glaucous; leaves mainly of leucocysts	3
2	Plants not glaucous; no leucocysts present	5
3	False mid-rib present	LEUCOPHANES* (p. 41)
3	No false mid-rib	4

4	Plant rigid; leaves in three rows	ARTHROCORMUS* (p. 42)
4	Plant lax; leaves spiral or in more than three rows	LEUCOBRYUM* (p. 39)
5	Basal hyaline sheath present on leaves	6
5	No basal hyaline sheath	9
6	Primary stem creeping	THYRIDIUM* (p. 62)
6	Primary stem erect	7
7	Gemmae present on upper leaves	8
7	No gemmae present on leaves	SYRRHOPODON* (p. 56)
8	Leaf margin narrowly bordered	CALYMPERIOPSIS* (p. 55)
8	Leaf margin not bordered	CALYMPERES* (p. 50)
9	Stem erect (acrocarpous moss)	10
9	Stem creeping (pleurocarpous moss)	18
10	Upper leaf cells prosenchymatous	11
10	Upper leaf cells parenchymatous	13
11	Leaf cells narrowly linear	MICRODUS* (p. 28)
11	Leaf cells elongated rhombic or rectangular	12
12	Alar cells present	CAMPYLOPUS* (p. 37)
12	No alar cells	BRYUM* (p. 77)
13	Primary stem creeping with erect branches	14
13	Primary stem erect	15
14	Leaves all alike; capsule on long seta	MACROMITRIUM* (p. 83)
14	Leaves dimorphous; capsule immersed	DESMOTHECA* (p. 82)
15	Leaf cells smooth	RHIZOGONIUM* (p. 78)
15	Leaf cells papillate or mamillate	16

16	Leaf lamina 2–3 layered	DIPHYSCIUM* (p. 79)
16	Leaf lamina one layered	17
17	Mid-rib rough on dorsal surface	SEMIBARBULA* (p. 73)
17	Mid-rib not rought on dorsal surface	HYDROGONIUM* (p. 74)
18	Leaf with prominent double mid-rib	CALLICOSTELLA* (p. 94)
18	Leaf without prominent double mid-rib	19
19	Leaf cells parenchymatous	20
19	Leaf cells prosenchymatous	24
20	Cells wide, hexagonal	DISTICHOPHYLLUM (p. 89)
20	Cells small	21
21	Paraphyllia absent	22
21	Paraphyllia present	PELEKIUM* (p. 96)
22	Leaves dimorphic; capsule immersed	DESMOTHECA* (p. 82)
22	Leaves all alike; capsule on long seta	23
23	Basal leaf cells elongate, incrassate	MACROMITRIUM* (p. 83)
23	Basal leaf cells not differentiated	PINNATELLA* (p. 88)
24	Alar cells absent	25
24	Alar cells present	27
25	Branch leaves complanate, not homotropous	26
25	Branch leaves radially arranged, not complanate; falcate and homotropous near tips	ECTROPOTHECIUM* (p. 97)
26	Regularly pinnate. Leaf with indistinct border	VESICULARIA* (p. 102)
26	Irregular branching. No border.	ISOPTERYGIUM* (p. 101)

27 Alar cells in one distinct row 28
27 Alar cells in several distinct rows 35

28 Cells smooth 29
28 Cells papillate 31

29 Cells thick-walled, margin always dentate .... TRISMEGISTIA* (p. 107)
29 Cells thin-walled, margin smooth or dentate 30

30 Apex forming an aristate point .... RHAPHIDOSTICHUM* (p. 112)
30 Apex blunt or with subulate point .... SEMATOPHYLLUM* (p. 111)

31 Cells thick-walled 32
31 Cells thin-walled 33

32 Cells elongate, rhombic .... ACANTHORRHYNCHIUM* (p. 103)
32 Cells linear .... TAXITHELIUM* (p. 105)

33 Cells papillate by projecting ends of cells .... MEIOTHECIUM* (p. 108)
33 Cells papillate over lumen 34

34 Seta scabrous above .... TRICHOSTELEUM* (p. 112)
34 Seta smooth .... WARBURGIELLA* (p. 111)

35 Complanate; secondary stems simple or irregular .... PLAGIOTHECIUM* (p. 97)
35 Not complanate; secondary stems usually pinnate .... ECTROPOTHECIUM* (p. 97)

BOG MOSS FAMILY

(Sphagnaceae)

Bog mosses are erect whitish green or pinkish plants which grow in dense cushions or form large tufts in wet acid conditions. In the tropics they occur on the tops of mountains and in some tropical swamp forests in the lowlands. Bog mosses are so different from the rest of the mosses that some scientists put them into an entirely different group of plants. Although they are responsible for peat formation in temperate regions, most tropical peat is formed from higher plants, not from Bog Mosses which are relatively rare. They appear to have been originally temperate plants which have, in past ages, spread to the tropics along the mountain ranges.

Bog mosses when dry can absorb many times their own weight in water due to the presence of dead cells in the outer parts of the stem and in their leaves. The main stem is upright and bears branches in groups or whorls. Some of the branches act as "wick branches" to draw up water from below, while the short ones at the top of the stem form an attractive flower-like head. Both the main stem and the numerous branches are tightly clothed with nerveless leaves. The shape of the leaves on the main stem differs from that on the branches; but all leaves consist of a network of extremely narrow green cells (*chlorocysts*) alternating with large dead often spirally-thickened cells (*hyalocysts*). Hyalocysts bear a number of large pores. When dry they are full of air so the plant has a whitish appearance; but when they are wet and full of water, the plant is bright green. Dry branches have a tremendous absorptive capactiy. They also act as wicks. If a long dry stem is placed in a tumbler of water so that part hangs over the side, it will drain off the water in the tumbler until it is empty. In the natural state the lower parts of the stem of a cushion of Bog Moss consists of dead and decaying branches. These form a wet soggy mass of very acid water. In Klang Swamp Forest, West Malaysia, the pH of the water may be as low as 3.8. Bog Moss does not normally bear rhizoids except for very young plants formed from spores. Unlike other mosses, the protonema is thallose.

Capsules of Bog Mosses, while numerous in temperate regions, are extremely rare or unknown for most tropical species. A Bog Moss capsule is a dark brown spherical structure which never has any chloroplasts. Its seta is very short, but the capsule is raised into the air by the outgrowth of a stalk from the head of the female plant. Bog Moss capsules have a lid but lack peristome teeth. In temperate regions, loud "pops" can be heard during a sunny day in a *Sphagnum* bog which are caused by the explosive mechanism of operculum release.

There are about seven species of *Sphagnum* (Bog Moss) in Malaysia. The Flaccid-Leaved Bog Moss (*Sphagnum flaccidifolium*) is found in ditches in the swamp forest at Klang. Most of the other species are confined to mountains above 900 m.

FLACCID-LEAVED BOG MOSS

Sphagnum flaccidifolium (Fig. 1)

The Flaccid-Leaved Bog Moss is an aquatic plant with long trailing stems, found in acid water. The stem is about 1 mm in diameter and bears small triangular leaves, about 1.6 × 1.0 mm. The branch leaves are exceedingly long, lanceolate, and about 3.0 – 5.0 mm long and 0.6 mm wide.

Localities: Telok Forest Reserve, Klang; Telok Panglima Garang. Also found in Sumatra.

JUNGHUHN'S BOG MOSS

Spagnum junghuhnianum (Fig. 2)

The Junghuhn's Bog Moss is a robust plant, often pinkish, reddish, or violet in colour. The main stem has thickened woody cells in the centre, surrounded by two or three layers of dead absorptive cells. The main stem leaves are triangular or tongue-shaped, about 2 × 1 mm, with the tip cut-off (*truncate*) and the sides of the upper part turned in. Branches arise in clusters of threes. The central branch of each cluster forms a drooping wick branch, while the other two are horizontal and more limited in growth. The branch leaves are quite large, 2 × 0.8 mm. They are entirely different in shape from the stem leaves, being oval below and drawn out to a cut-off tip.

Localities: Widespread in the mountains of Malaysia, usually above 900 m. Kedah Peak, Gunong Sitong, Gunong Ledang (Mount Ophir), Gunong Batu Brinchang, Gunong Tahan, Cameron Highlands. Also found in Japan, Taiwan, Himalayas, Indonesia.

◆ ◆ ◆

FORK AND SWAN MOSS FAMILY

(Dicranaceae)

Mosses in the Fork and Swan Moss family are extremely variable in size but all are upright with sparse branches, or branch by innovation. The leaves are narrow, long pointed, spirally arranged, with a distinct nerve running to or beyond the apex. In some genera all the leaves are characteristically turned to one side (*falcate*). Lamina cells are often very

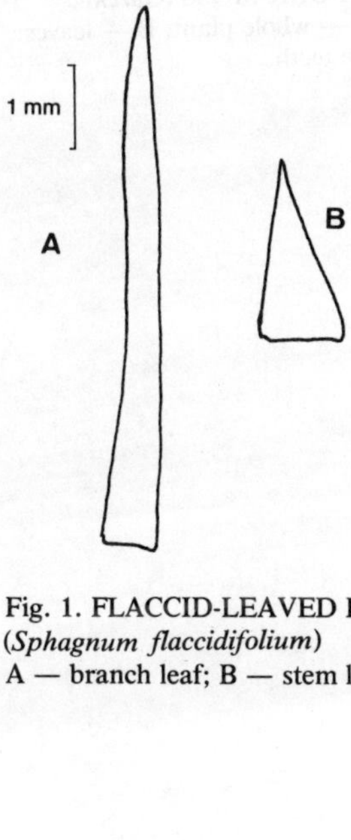

Fig. 1. FLACCID-LEAVED BOG MOSS
(*Sphagnum flaccidifolium*)
A — branch leaf; B — stem leaf.

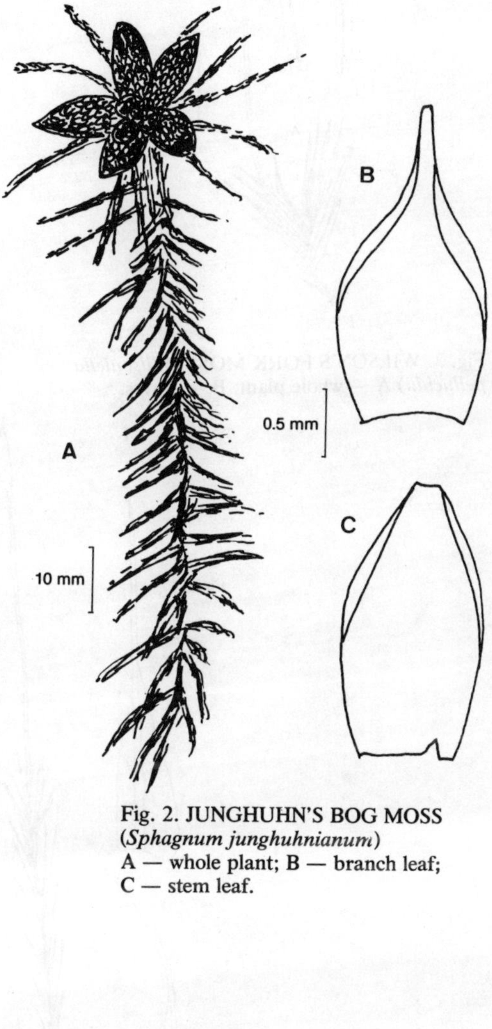

Fig. 2. JUNGHUHN'S BOG MOSS
(*Sphagnum junghuhnianum*)
A — whole plant; B — branch leaf;
C — stem leaf.

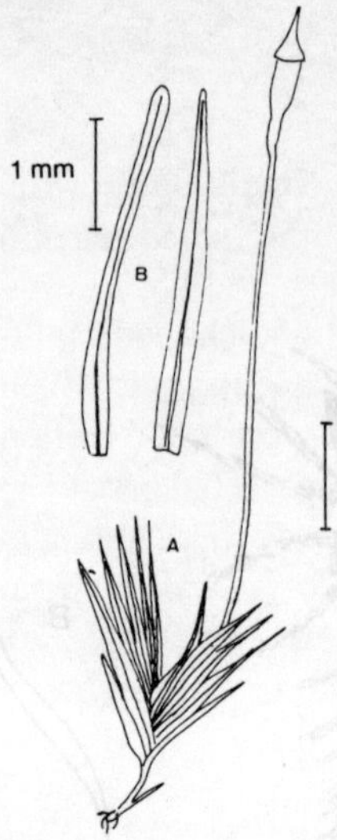

Fig. 3. WILSON'S FORK MOSS (*Wilsoniella pellucida*) A — whole plant; B — leaves.

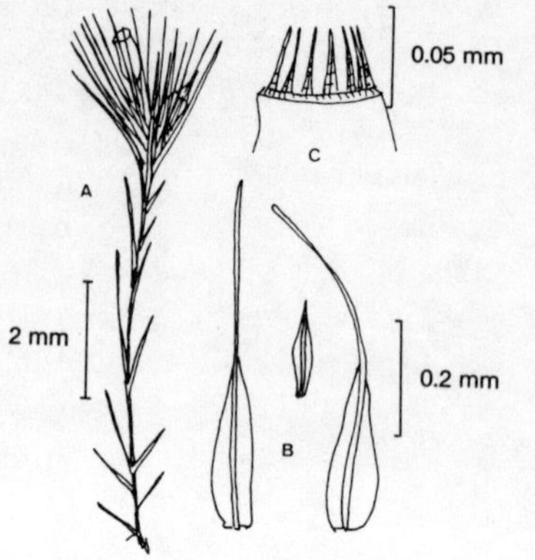

Fig. 4. VASE FORK MOSS (*Garckea phascoides*) A — whole plant; B — leaves; C — peristome teeth.

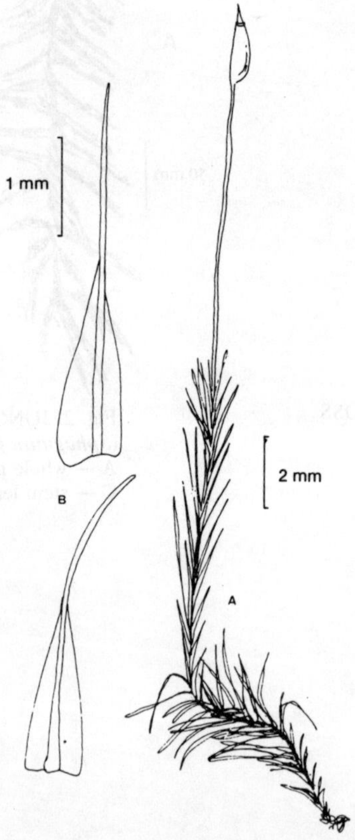

Fig. 5. DIFFICULT TWO-HAIRED FORK MOSS (*Ditrichum difficile*) A — whole plant; B — leaves.

small, and rectangular in shape. Some genera have small outgrowths (*auricles*) at the basal corners of the leaves. Others may have small patches of bright yellow or orange cells on either side of the base.

In most cases the calyptra is smooth, narrow, and hood-shaped, splitting down one side. The capsule is usually elevated on a long seta, which in Swan mosses is curved like a swan's neck. The operculum is often prolonged into a long point which may be asymmetrical. The peristome consists of a single row of 16 teeth, each of which is divided into two forks. The scientific name of the family, Dicranaceae, is derived from the Greek word *dikranon* meaning a two-pronged fork, and refers to the peristome teeth. The teeth are reddish in colour and are thickened by horizontal bars. This type of peristome is known as *dicranoid*.

WILSON'S FORK MOSS

Wilsoniella pellucida (Fig. 3)

The Wilson's Fork Moss is a small, pale, bluish-green moss found growing on rocks in caves. The stem is short, upright, and about 5 mm tall. It bears a few red rhizoids at the base. The lowermost part of the stem may be prostrate or nearly so, and one or more true branches may occur above. The leaves are narrow, tongued-shaped, erect-spreading away from the stem, and only slightly reduced in size at the base. The laminal cells are rhombic or rarely hexagonal and not much differentiated at the base. The seta is terminal on the stem, about 1.5 cm high, upright, and dull brown in colour. The capsule is long, cylindrical, with a narrow, partly-defined neck. The operculum has a conical base and a very long tip. The calyptra is hood-shaped and splits down one side. The peristome consists of a single row of 16 teeth, attached to a membrane at the base, with each tooth split into two long filiform arms which are deeply inserted under the mouth. The teeth, which are transversely thickened and very papillose, are extremely fragile. They often break off half-way down.

Localities: Batu Caves, Selangor. Found also in Indonesia and Sri Lanka.

VASE FORK MOSS

Garckea phascoides (Fig. 4)

The Vase Fork Moss is a palish green moss, about 1 cm high. It has a thin, erect stem with rhizoids at the base. The lower leaves are smaller than the upper ones and, when dry, are closely pressed to the stem. The upper leaves are much larger and form a distinct head. Each leaf is widened at the base. It has a prominent nerve which extends into the long hair-like tip at the apex. Lamina cells are longer than wide with oblique cross walls, except for two prominent patches of wide yellowish cells at either side of the base.

The capsule may be several in a head at the top of the stem. The seta is very short, with the upper leaves surrounding the capsules. Each capsule is vase-shaped. The operculum is fairly long, often with a bent tip. The 16 peristome teeth are deeply inserted and tend to stick together, leaving gaps between them. When young the capsule is covered with a bell-shaped calyptra which does not split down the side.

Localities: The Vase Fork Moss is only found in the north-west of Peninsular Malaysia and has been recorded from Penang Hill, Kedah Peak, and in Perlis. It occurs on wet banks by the road side. It is widely distributed in Indonesia, Thailand, and the Andaman Islands.

DIFFICULT TWO-HAIRED FORK MOSS

Ditrichum difficile (Fig. 5)

The Difficult Two-Haired Fork Moss is a tufted moss of golden green colour, less than 1.5 cm tall. Its base is sometimes prostrate but it soon turns upwards to bear long drawn-out pointed leaves. The nerve is very prominent and is strongly excurrent. The leaves are flexuose, when dry forming a canal-like channel at the tip. There is no differentiation into special groups of basal cells. The laminal cells are small and rectangular.

The seta is borne terminally, is as long as or even longer than the stem, and red or yellow in colour. It is sometimes slightly wavy. The capsule is strongly asymmetrical and swollen on one side. The annulus is distinct, with the operculum rather small and asymmetrically placed. The wall of the mature capsule has longitudinal furrows. The peristome consists of 16 forked teeth, joined to a basal membrane.

Localities: This is found as a ground or rock moss in the mountains at Gunong Keledang, Perak.

*MIQUEL'S SMALL-TOOTHED FORK MOSS

Microdus miquelianus (Fig. 6)

The Miquel's Small-Toothed Fork Moss is a very common but tiny moss, only 2 — 5 mm high. It grows in loose tufts on bare ground in the sun and is nearly always fertile. However it is so small that it is easily overlooked. The rhizoids are short, consisting of many cells with oblique cell walls, but are rarely branched. Some rhizoids are coiled at the base. The stem is unbranched or has a single innovation above. The leaves are a long triangular shape, gradually narrowing from the base. They are rather firmly attached to the stem and difficult to remove for microscopic examination. The lower leaves are only one-quarter of the size of the upper ones. The upper leaves have a more expanded base. The nerve is prominent and extends to the end of the leaf. The margin may be smooth or very slightly toothed. The upper part of the leaf is often folded inwards above when dry, and may be twisted. Most of the lamina consists of small long cells, but the base has a few rounded wider cells.

The seta is terminal, 1 cm long, often orange below and yellow above, twisted when dry, and with a prominent vaginula at the base. The capsule is upright, cylindrical, elongated at the base, and exceedingly twisted when dry. The capsule wall has large irregular stomates present, visible under the microscope. The operculum is long, pointed, with spirally arranged cells with papillate projections. The teeth are in a single row, short, stumpy, irregular, and incomplete.

Localities: Very common on the ground throughout the country. Recorded from Penang; Tapah; Perak; Selangor; Perhentian Tingii, Negri Sembilan; Tana Rata, Cameron Highlands; and Singapore.

COMPRESSED FORK MOSS

Dicranella coarctata (Fig. 7)

The Compressed Fork Moss forms small loose tufts less than 1 cm high. The stem is unbranched with only a few leaves near the base which become more numerous towards the tip. The leaves have sheathing bases which turn abruptly away from the stem at an angle of about 60° to form a long stiff extended apical part. Laminal cells are retangular or narrowed, irregular in size, and somewhat thick-walled. The lowermost cells are brightly coloured. The nerve is conspicuous but narrow, only occupying one-sixth of the base in width.

The seta is prominent and upright, about 1.5 cm tall, and pale yellow or red in colour. The capsule is upright, ovoid, and pale brown when ripe. It does not bear stomates. The mouth of the capsule has a prominent rolled-over annulus which gives a constriction between the operculum and the capsule. The peristome teeth are inserted well under the mouth. Each tooth is split into two long forks. The operculum has a conical base with frilly margin. It is lengthened to a long point, almost as long as the capsule.

Localities: This is a ground moss common in the mountains at Fraser's Hill and Kedah Peak at about 1,000 m. It is widely distributed in South-east Asia, including Indonesia.

TRUE-BRANCHED SWAN MOSS

Campylopodium euphorocladum (Fig. 8)

The True-Branched Swan Moss is an upright moss forming thick tufts which may become matted with earth. Individual plants may be 15 cm tall. They are dark velvety green at the base and paler above. One or two true branches may be present. The leaves are large, about 5 mm long, and stand out stiffly from all sides of the stem. The lanceolate base narrows to form a long rolled awl-shaped upper part, with a strongly incurved margin which may be smooth or slightly toothed. The nerve is wide and occupies one-third of the base in width. It is convex on both sides of the leaf, and made of long cells with curved walls on the outside and thickened inside. The lamina cells are smooth and regular rectangular, with the lowermost forming two bright yellow patches on either side of the nerve at the base. The lower part of the leaf has a short border of hyaline cells, disappearing above.

The seta is about 5 mm long, bent over like a swan's neck when young (*cygneous*). Mature capsules are more or less upright, with a short swollen neck. Stomates may be seen on the capsule wall under the microscope. The capsule, which is more or less ovoid in shape apart from the neck, dries in longitudinal furrows. The operculum is short-pointed and covered with a small hood-like calyptra with a single split when young. The peristome is dicranoid.

Localities: This is a ground moss found at Kedah Peak, Fraser's Hill, Penang Hill, in Selangor, and the Cameron Highlands. It is widely distributed throughout Indonesia, Sri Lanka, West Australia, the Philippines, and Taiwan.

BRAUNFELS' FORK MOSS

Braunfelsia discranoides (Fig. 9)

The Braunfels' Fork Moss is a stiff upright moss about 4 cm tall. The stem is very brittle, and is either single or branched once by innovation. The leaves are strongly turned to one side of the stem (*falcate*), and are concave on the ventral side. The leaf base is lanceolate, with long inrolled apical part above. The nerve is not very prominent, being only a few cells wide, but it just reaches the tip. Laminal cells are narrow, except for three rows of brown-coloured thick-walled alar cells at the two leaf corners at the base.

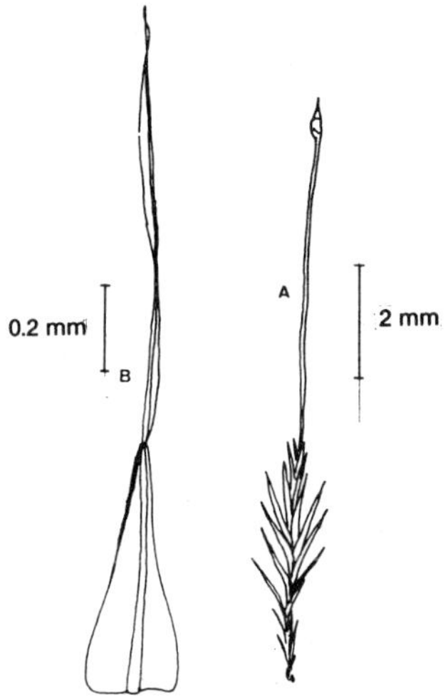

Fig. 6. MIQUEL'S SMALL-TOOTHED FORK MOSS (*Microdus miquelianus*) A — whole plant; B — leaf.

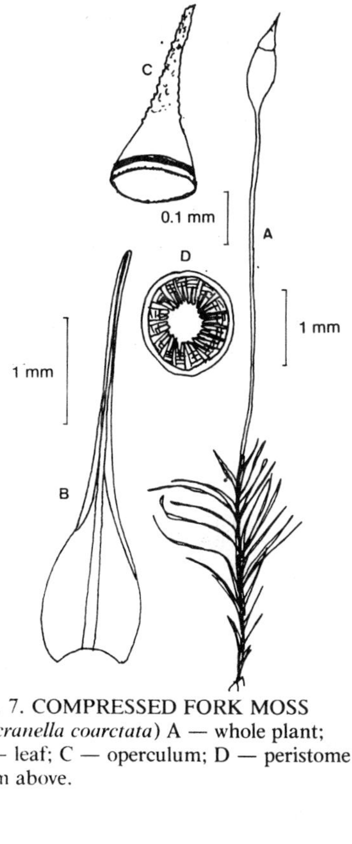

Fig. 7. COMPRESSED FORK MOSS (*Dicranella coarctata*) A — whole plant; B — leaf; C — operculum; D — peristome from above.

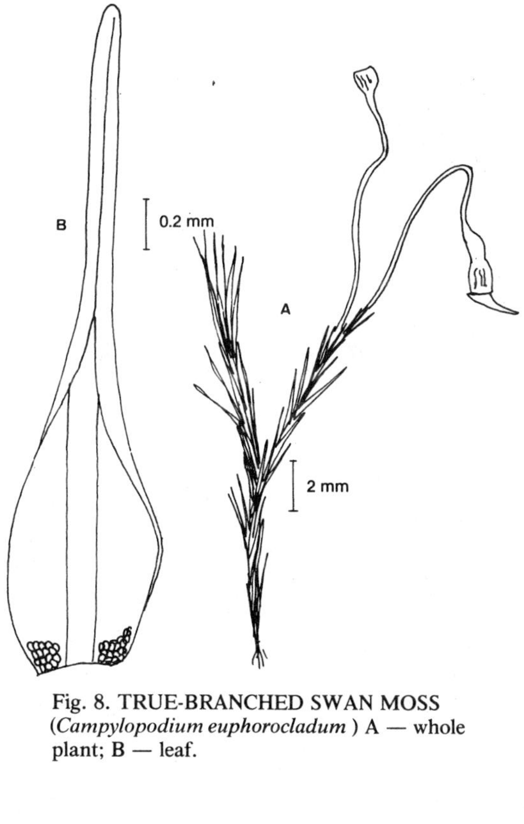

Fig. 8. TRUE-BRANCHED SWAN MOSS (*Campylopodium euphorocladum*) A — whole plant; B — leaf.

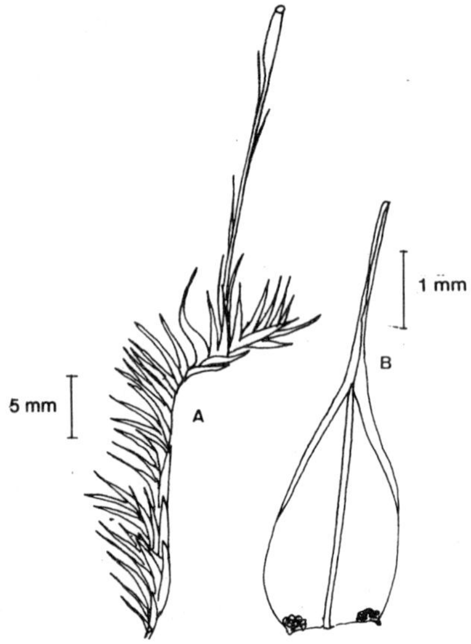

Fig. 9. BRAUNFELS' FORK MOSS (*Braunfelsia dicranoides*) A — whole plant; B — leaf.

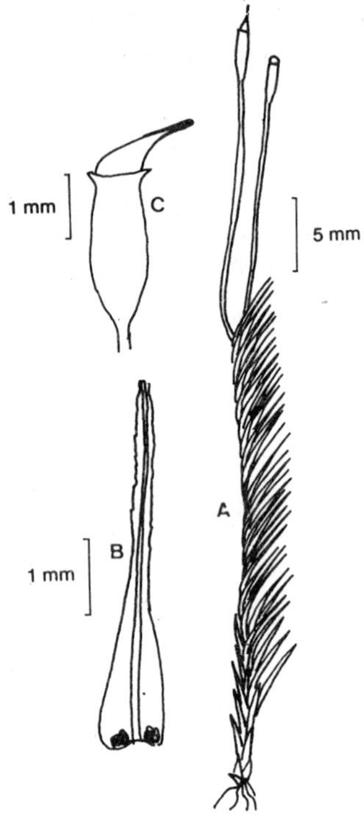

Fig. 11. SIMILAR FORK MOSS (*Dicranoloma assimile*) A — whole plant B — leaf; C — capsule.

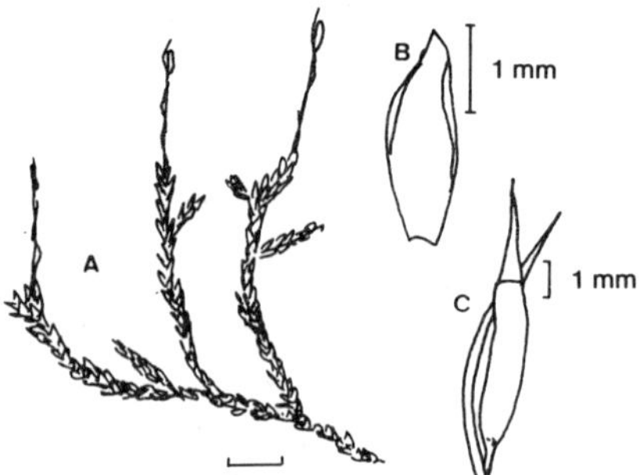

Fig. 10. BRAUNFELS' NERVELESS MOSS (*Braunfelsia enervis*) A — whole plant; B — leaf; C — capsule with tip of one perichaetial leaf.

The seta is brown in colour, borne terminally on the stem, and often ensheathed by the hair-like perichaetial leaves. The capsule is long and narrow, cylindrical and upright. The neck is not much differentiated, but the surfaces are faintly ridged when dry. The annulus is not very prominent. Unlike most members of this genus, the peristome is completely absent. The operculum has a long beak.

Localities: This species is only found on Gunong Tahan above 2,000 m.

BRAUNFELS' NERVELESS MOSS

Braunfelsia enervis (Fig. 10)

The Braunfels' Nerveless Moss is a robust plant growing in dense tufts, with a few rhizoids at the base. It may be several times branched. Leaves are crowded, spreading in all directions, completely nerveless, and twisted when dry. Perichaetial leaves have long tips coiled round the seta.

The seta is long but its structure is hidden by the enveloping perichaetial leaves. The capsule is somewhat swollen at the base. There is a distinct annulus but no peristome teeth. The operculum is long and pointed.

Localities: Mountains over 1,000 m on Fraser's Hill and Gunong Tahan. It is also found in Java.

SIMILAR FORK MOSS

Dicranoloma assimile (Fig. 11)

The Similar Fork Moss is a stiff upright plant, pale green in colour, which forms loose tufts about 4 cm high. Individual plants are rarely branched. The leaves are prominently turned to one side of the stem, and are awl-shaped, with a long tapering tip, toothed above. When dry the leaves curl into a hollow tube. The nerve is distinct and consists of thickened rectangular cells. It is about four cells in width. The lamina is made of thin-walled rectangular cells above, with a scarcely distinct border. There is a conspicuous yellow patch of cells on either side of the base of the leaf, usually not reaching the mid-rib. These yellow cells are rounded and thick-walled.

The capsule is red-brown, erect on a long seta about 2 cm long. It is urn-shaped. The operculum has a long point, often turned to one side. The peristome teeth are dicranoid, split two-thirds down into two arms, and distinctly thickened with reddish brown horizontal bars. The spores are spherical and bright yellow in colour.

Localities: It is normally found in mountains above 1,000 m. In Malaysia it occurs on Gunong Tahan, Gunong Hijau, and in the Semankok Pass. It is also found in Indonesia and the Philippines.

TENDER WHITE-BORDERED FORK MOSS

Leucoloma molle (Fig. 12)

The Tender White-Bordered Fork Moss is found in loose irregular tufts about 4–6 cm tall. It is pale green when living, brown when dry. The brown stem is slender, and has both true branching as well as branching by innovation. The most remarkable feature is the partially bald nature of parts of the stems caused by the ready falling off of leaves. The remaining leaves are turned to one side (*falcate*), stand at a fair angle to the stem, and are concave with the upper part rolled into a hollow tube. The leaves are roughly triangular in shape, with a long excurrent nerve. At the base of the leaf there are two special groups of rounded rectangular, yellow-coloured cells. Other laminal cells are small, round, green, and papillose. The leaves are distinctly bordered, with several rows of narrow, whitish, hyaline cells, beneath which lie a second "margin" of thick-walled cells with pointed ends.

At first sight the capsule appears to be borne laterally, with several formed by a single plant. In reality each capsule is at the end of a short lateral shoot. The seta is quite short, only 2 mm long. The capsule is also short, without a neck and with strong furrows when dry. The operculum is conical, with a long asymmetrical tip. The peristome is single, with 16 peristome teeth, each divided almost to the base into two long papillose arms. The calyptra is hood-shaped.

Localities: Found on elevated ground at Penang Hill, Bukit Hitam, the Dindings, and Gunong Tahan. This species is widely distributed in Indonesia, Sri Lanka, Hong Kong, the Philippines, Taiwan, and Japan. There are several varieties.

HAIRY SWAN MOSS

Campylopus comosus (Fig. 13)

The Hairy Swan Moss forms loose, pale green to golden tufts, with red rhizoids below. The stem is single or with true branching at the extreme base. It is about 2–6 cm high, very thick and wiry, and brown or red in colour. The leaves are sparse on the lower part of the stem where they tend to cling to the stem throughout their length. On the upper part they form a thick comal tuft, with the upper part of each leaf strongly bent away from the stem and somewhat curved to one side. The tip of

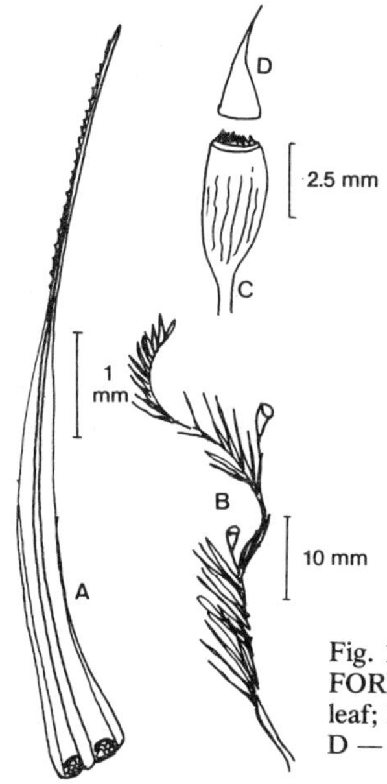

Fig. 12. TENDER WHITE-BORDERED FORK MOSS (*Leucoloma molle*) A — leaf; B — whole plant; C — capsule; D — operculum.

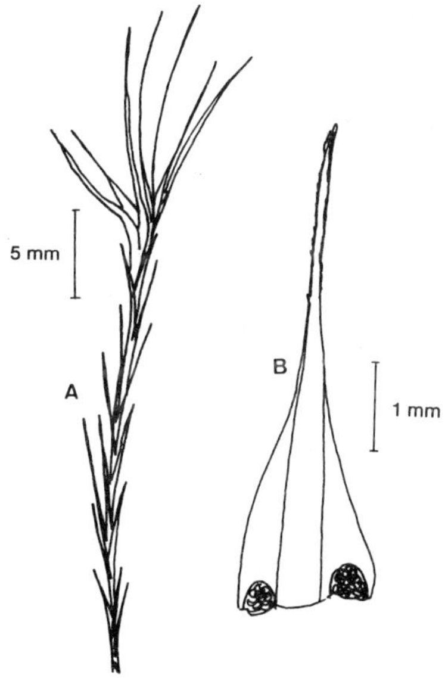

Fig. 13. HAIRY SWAN MOSS (*Campylopus comosus*) A — whole plant; B — leaf.

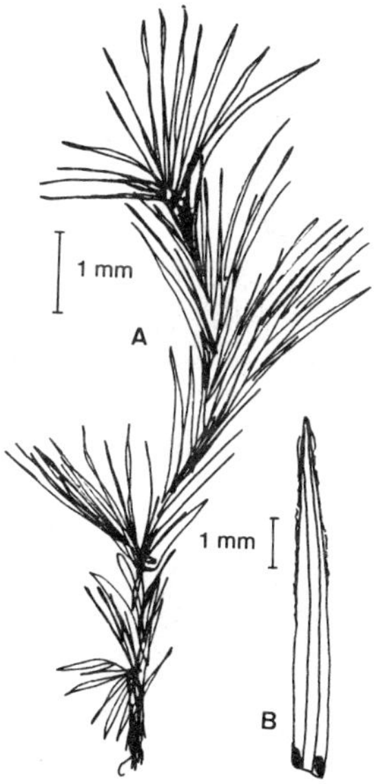

Fig. 14. TOOTHED SWAN MOSS (*Campylopus serratus*) A — whole plant; B — leaf.

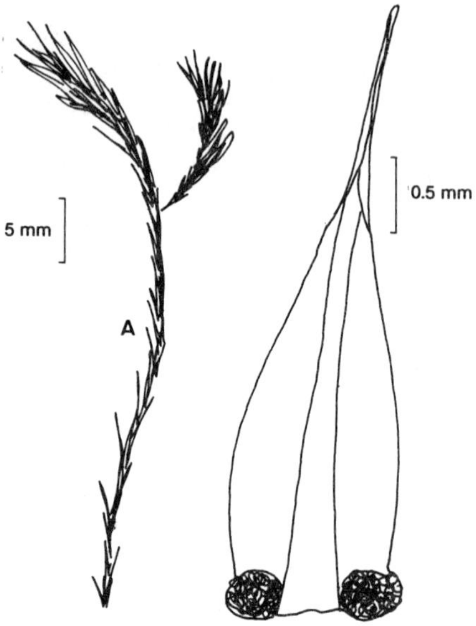

Fig. 15. THWAITES' SWAN MOSS (*Campylopus thwaitesii*) A — whole plant; B — leaf.

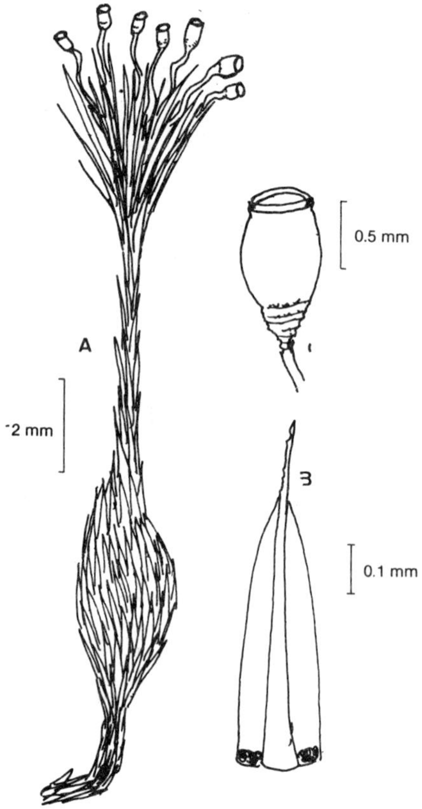

Fig. 16. RICHARDS' SWAN MOSS (*Campylopus richardii*) A — whole plant; B — leaf; C — capsule.

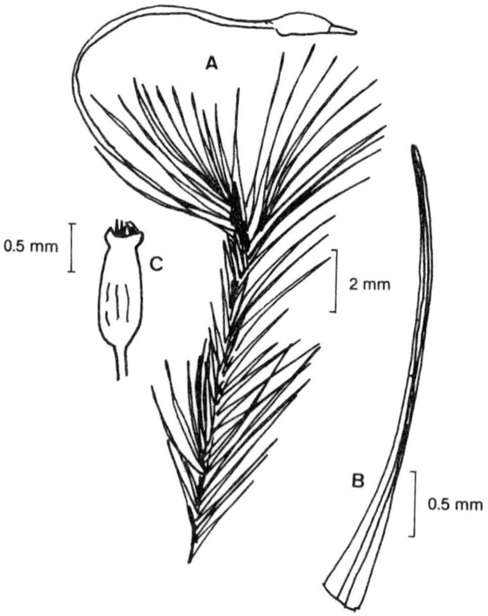

Fig. 17. SHINING FORK MOSS (*Dicranodontium nitidum*) A — whole plant; B — leaf; C — capsule.

the leaf is strongly doubled-toothed. The nerve occupies about one-third of the width of the leaf below. The narrow lamina curls over the nerve almost to the tip. Two groups of red or yellow, hyaline, 4–6 sided cells are found at the base. Ordinary laminal cells are elongated rhombic above, rectangular below. Outer cells near the margin are narrower, but do not form a distinct border.

This moss is usually sterile; it rarely bears a seta 1 cm high and yellow in colour. It bears the capsule at about 90° angle to the seta. There is a distinct neck and the capsule is much furrowed above. The peristome teeth are dicranoid, 16 in number, red, papillose, and deeply inserted

Localities: Penang Hill, Kedah Peak, Fraser's Hill. Also found in Java, South America, and New Guinea.

*TOOTHED SWAN MOSS

Campylopus serratus (Fig. 14)

The Toothed Swan Moss is a large, robust, upright species which forms patches about 1.4 cm tall. The stem shows a tendency to zig-zag, and may be branched by innovation, just below the top. The leaves are spreading, rigid, and olive-green in colour, and even darker at the base. The lower leaves are distinctly blackish and are smaller than the upper leaves which are about 5 × 1 mm. The nerve is wide, occupying at least one-third of the width of the base. There is a patch of irregular thickened pigmented cells on either side of the nerve at the base. Upper lamina cells are thick walled, narrower and longer than lower cells. Leaf bases are clothed with red or yellow branched paraphyllia. The leaf tip is strongly toothed, with an excurrent nerve. No capsule has been seen.

Localities: on the ground — Kedah Peak; Penang Hill; Taiping; Ginting Simpah; Bukit Timah and Botanic Gardens, Singapore.

THWAITES' SWAN MOSS

Campylopus thwaitesii (Fig. 15)

The Thwaites' Swan Moss is a moderately sized moss growing in untidy loose tufts, 3–6 cm high. The basal part is dark green to brownish black in colour, while the top is a beautiful golden yellow colour. Branching is rare and by innovation. The stem is very thin throughout. The leaves tend to cling to the stem, being closely adpressed when dry, so that it is difficult to get them off for examination. When wet they stand off from the stem at a slight angle. The leaves are about 5 mm long, ovate below, but with a long awl-shaped upper part. The leaf features are very difficult to see since each leaf dries to form a hollow tube. In the pale-coloured leaves near the tip, the nerve is not very distinct since its colour is identical with that of the lamina. The older leaves are much darker and the nerve is much more prominent, occupying one-third of the total width of the base and gradually tapering to form a protruding tip at the apex. This nerve is made of long, narrow, thick-walled rectangular cells, and is extremely spiny above. The ordinary lamina cells are small, rectangular, and similar from top to bottom of the leaf. Basal cells form two patches of yellow, brown, or sometimes colourless angular cells. This basal part is two cells thick, and is very prominent in the detached leaf. No capsules have been seen.

Localities: This is a ground and rock moss found on Gunong Padang, Trengannu, at about 1,300 m. It is also found in Java and Sri Lanka.

RICHARDS' SWAN MOSS

Campylopus richardii (Fig. 16)

The Richards' Swan Moss is quite an extraordinary moss growing in rigid tufts about 3.5 cm high. The basal part of the stems grows along the ground and thickens considerably when turning upwards, whereupon it bears one or more fertile branches. The fertile stem is slender and bears closely adpressed leaves, ending in a head of normal and perichaetial leaves surrounding the archegonia in the female. In the male plants the leaves are less adpressed below the male head. Normal leaves are about 1 mm long, with an ovate base and a long tip consisting of a long, spiny, protruding nerve. Perichaetial leaves are very narrow, and the tip is long excurrent to three times the length of the leaf.

Capsules are borne several on a female head. They have long setae, bent like a swan's neck. Their shape is ovoid with a wide mouth. The neck is very rough, bearing several rows of warts near its base. The peristome consists of 16 forked teeth.

Localities: Found in the Cameron Highlands at 1,500 m and above.

SHINING FORK MOSS

Dicranodontium nitidum (Fig. 17)

The Shining Fork Moss forms tall, loose, golden-green tufts, 1—10 cm high. The tufts are distinctly shiny. Red rhizoids are found at the base. The stem is upright, often unbranched, but usually bent and apparently articulate. The leaves are triangular, but long extended above. The lamina does not reach the tip which may be more or less tubular or flat. The nerve may be warty above. Lamina cells are long, with oblique cross walls. The border is very prominent, consisting of very narrow cells, and is about 6 cells wide. The nerve occupies about one-fifth of the base in width. Leaves are usually strongly falcate.

The seta is long and yellow, strongly curved, and about 1 cm long. The capsule is cylindrical, without a prominent neck. The operculum has a long tip. The calyptra is hood-like, split down one side. The dicranoid peristome is inserted below the mouth.

Localities: Above 1,000 m on Gunong Sitong, Maxwell's Hill, and Gunong Tahan. Widely distributed in Indonesia, Sri Lanka, and the Philippines.

WHITE MOSS FAMILY

(Leucobryaceae)

White mosses are very common in the tropics, especially in the lowlands. They are recognized by their glaucous green, whitish colour. They grow in cushion-like tufts and are capable of absorbing considerable quantities of water to form a wet spongy mass. Each plant of a tuft is usually quite sturdy with a thick main stem which is either unbranched or has one or two true branches. The stem is clothed with many rows of leaves which may be turned to one side (*falcate*). Under the microscope the leaves are quite different from those of other mosses, since they are composed almost entirely of a thick nerve which is many cells wide and several cells thick. Most of the cells are large, inflated, dead, colourless cells with pores on their inner walls; but a cross-section of the leaf will show one or more rows of very small, narrow living green cells sandwiched between the dead cells. The true lamina is only one cell thick. It is very narrow and consists entirely of dead cells.

In many of our local species capsules are rarely found. They are usually oblong in shape, either erect or inclined at an angle to the seta. The peristome, which is somewhat variable, is always in a single row. The capsule wall has no stomates.

*HOLY WHITE MOSS

Leucobryum sanctum (Fig. 18)

The Holy White Moss is one of the commonest mosses in Singapore and Malaysia. Its name is derived from the first collection from a temple in Java. It is a moderate sized moss forming loose, whitish green tufts. Old stems trail on the ground, and may be as much as 10 cm long but are often much shorter. The leaves are somewhat concave and turned to one side of the stem. They are about 5 × 1 mm. Most of the leaf is made of a nerve of three layers of cells. The upper and lower layers are dead cells (*leucocysts*) while the middle layer consists of tiny green cells (*chlorocysts*). Leucocysts differ from the hyalocysts of *Sphagnum* since they do not have spiral thickening and the pores have different construction. This species has distinct auricles on either side of the base of the leaf.

The capsule is inclined on a shiny red seta. It is ovoid and green when young, but crumples or folds up on drying to give the widest portion at the mouth. The operculum is pointed and covered by a calyptra which splits down one side. Beneath the operculum is the red-brown peristome consisting of a single row of 16 dicranoid peristome teeth. Each tooth is split to two-thirds to the base to form separate arms, which have horizontal bars and numerous papillae.

Localities: The Holy White Moss is found throughout the area ususally in shady situations on soil or rotten wood. In Singapore it is common on the paths on Bukit Timah. It is found throughout Southeast Asia, in the Pacific Islands, and in Australia.

*JAVANESE WHITE MOSS

Leucobryum javense

The Javanese White Moss is a large robust version of the Holy White Moss. It grows in loose tufts 6—12 cm high. The leaves are crowded, lanceolate, falcate, and long tipped, about 12 × 2 mm. The tip of the leaf forms a hollow hooded apex which is rough dorsally. Most of the leaf consists of a nerve which is made of many layers of dead cells with a single central layer of tiny green cells. The lamina is very narrow, only four to six cells wide.

The seta is about 2.5 cm tall. The capsule is short-oblong in shape, inclined with a swollen neck, and wrinkled when dry. The operculum is conical in shape.

Localities: This is rarer than the Holy White Moss but is found in similar localities in forest on the ground, often at somewhat higher elevations. It has been recorded throughout Malaysia and from Singapore. It is also found in Indonesia and Thailand.

*HOOKED WHITE MOSS

Leucobryum aduncum (Fig. 19)

The Hooked White Moss is a small to medium sized plant forming whitish cushion-like tufts. The stems are sparingly branched and 1—5 cm tall. The leaves are shiny and iridescent when dry. Each leaf is about 3 mm long and 0.5 mm wide, and is strongly falcate. The tip of each leaf is rough on the under (dorsal) side. The capsule is pendulous, and borne on a thin brown seta which is about 2 cm tall.

Localities: Recorded from Malacca and Singapore. Widely distributed in Sarawak. Indonesia, Sri Lanka, New Hebrides, Thailand, and Hong Kong.

*EIGHT-ROWED WHITE MOSS

Leucophanes octoblepharioides (Fig. 20)

The Eight-Rowed White Moss is a common species normally found on the bark of trees and forming short, dense, untidy tufts, 3 cm high. The tongue-shaped leaves are rather loose and spreading. Most of the leaf consists of nerve but there is a prominent band of thickened cells running down the centre of this, which forms a false nerve (*pseudonerve*). Each leaf is about 4 $\times$ 0.5 mm, with a apex which is bent back when dry.

The rough brown seta is about 4 cm long. The capsule is upright, cylindrical, and warm-brown when ripe. The operculum is conical. The peristome teeth are in a single row of 16 undivided teeth which are somewhat papillose. The young capsule is completely covered with a hood-shaped calyptra with a single lateral split.

Localities: Widespread in Singapore and Malaysia on the bark of trees — Singapore, Kuala Lumpur, Ulu Bendul, Penang, Taiping, and Bukit Panchor. Distributed in the Philippines, Indonesia, India, sri Sri Lanka, Thailand, Pacific Islands, and Australia.

*PURE WHITE MOSS

Leucophanes candidum

The Pure White Moss is a small to medium sized moss, consisting of loose whitish-green tufts, 2—6 cm high. The individual plants are upright and branched two or three times by true branching. The leaves are very thick and dense (especially in the variety called *densifolium*). They stand out from the stem and may be bent downwards from the tip. They are never turned to one side. Their shape is roughly ovate, with a narrow apex and a strongly excurrent tip. The true lamina is only a few cells wide and consists of elongated cells. The nerve is very wide and occupies most of the leaf. Under the microscope this is a whitish tissue about three cells thick. The living green cells (*chlorocysts*) are difficult to see except under high power. In the centre there is a line of thickened elongated cells, often brown or yellow in colour. These form the false nerve (*pseudonerve*).

The capsule in this species of moss is always upright, urn-shaped, and of a dark red colour, usually with a lighter band at the annulus. There is a single row of 16 peristome teeth which are rather massive structures, isolated from one another at the point of attachment and covered all over with fine papillae. They are not split (cf. Dicranaceae and *Leucobryum*) and therefore cannot be described as dicranoid. The mouth is wide and covered with a cap-shaped operculum with a very long tip. The red seta is about 2 cm long.

Localities: Widely distributed in Singapore and Malaysia as an epiphyte on the lower part of tree trunks, on rotting wood, and occasionally on the ground. Also distributed throughout Southeast Asia and in Australia and the Pacific Islands.

*NARROW-LEAVED WHITE MOSS

Leucophanes albescens (Fig. 21)

The Narrow-Leaved White Moss species consists of tall, untidy plants forming medium-sized loose tufts to 6 cm high. The leaves are covered above and erect-spreading, rather narrow, about 3 × 0.3 mm, lanceolate in shape. They tend to be folded down the pseudonerve and undulate when dry. The leaf tip is distinctly toothed, with the teeth sometimes in pairs. The nerve is wide and consists of many layers of cells. The lamina consists of only a few rows of rectangular cells at the base and disappears above. The capsule has not been observed.

Localities: This moss has a curious distribution in Kedah; Trengganu; and Pulau Ubin, Singapore. It is probably much commoner. It is also found in Thailand.

*JOINTED WHITE MOSS

Arthrocormus schimperi (Fig. 22)

The Jointed White Moss is a remarkable moss which can be found growing in large cushions or patches by itself, or as individual, isolated plants mixed with other mosses, especially on tree trunks or palm stems in shady forests. Plants are extremely stiff and rigid, and scarcely look like living plants at all. In cushions they grow extremely closely together. The colour is the typical whitish bluish green associated with the White Moss family. When dry this species may become a shiny silvery colour. Both stem and leaves are very brittle, particularly when dry, and are

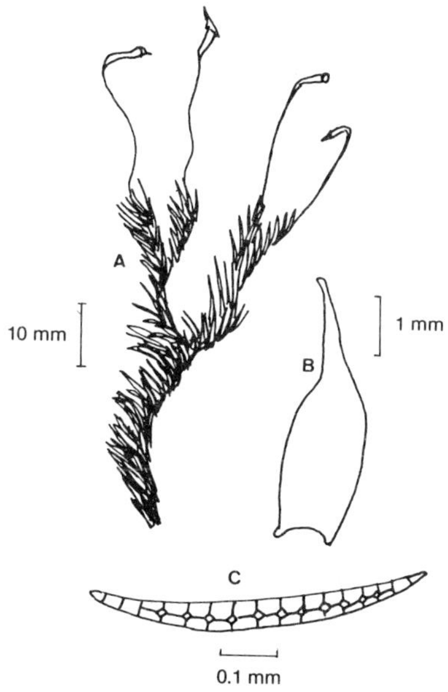

Fig. 18. HOLY WHITE MOSS (*Leucobryum sanctum*) A — whole plant; B — leaf; C — transverse section of a leaf.

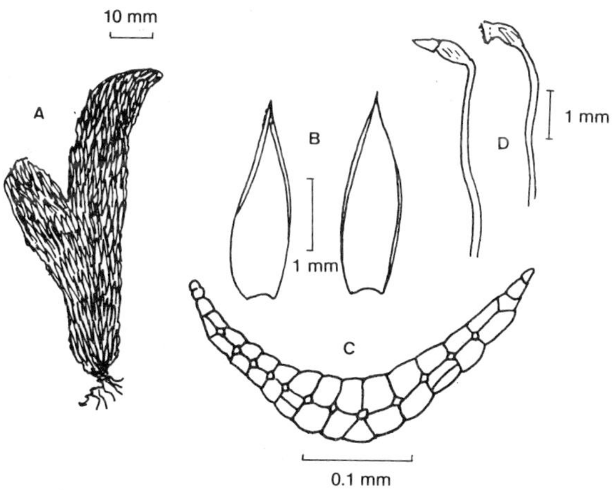

Fig. 19. HOOKED WHITE MOSS (*Leucobryum aduncum*) A — whole plant; B — leaves; C — transverse section of a leaf; D — capsules.

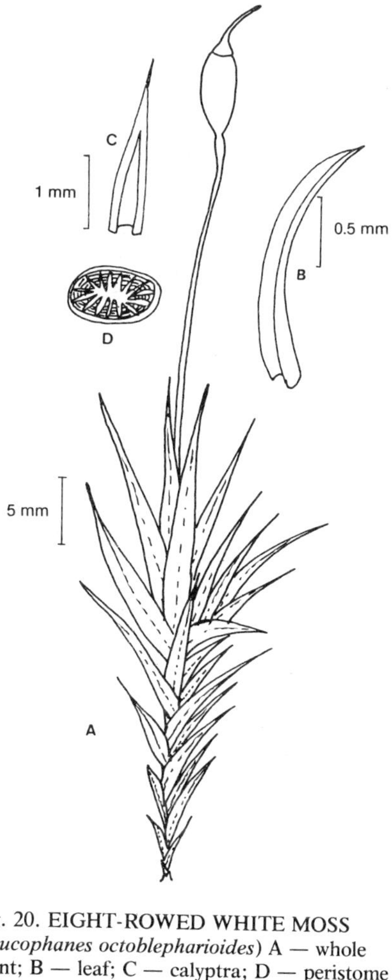

Fig. 20. EIGHT-ROWED WHITE MOSS (*Leucophanes octoblepharioides*) A — whole plant; B — leaf; C — calyptra; D — peristome teeth.

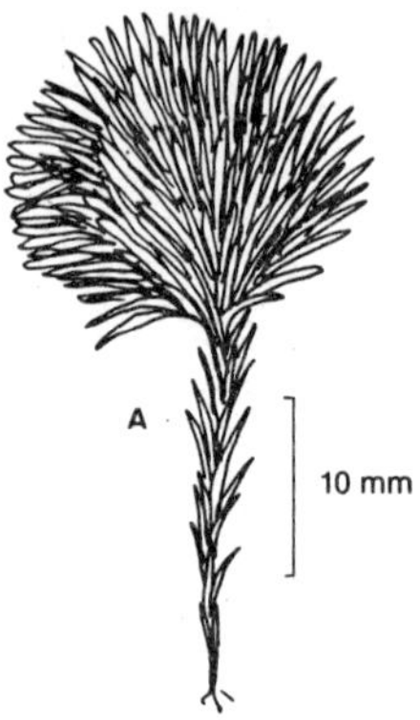

Fig. 21. NARROW-LEAVED WHITE MOSS (*Leucophanes albescens*) A — whole plant.

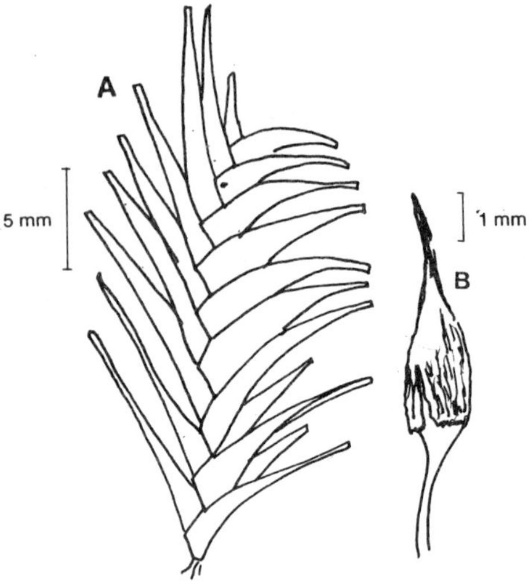

Fig. 22. JOINTED WHITE MOSS (*Arthrocormus schimperi*) A — whole plant; B — capsule with calyptra.

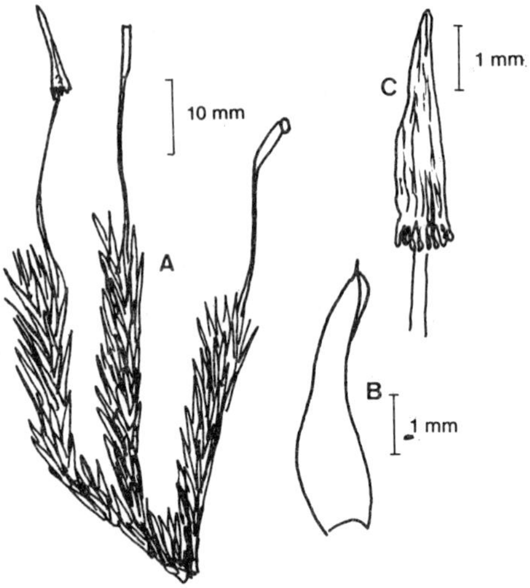

Fig. 23. POINTED SPLIT-CAP WHITE MOSS (*Schistomitrium apiculatum*) A — whole plant; B — leaf; C — capsule with calyptra.

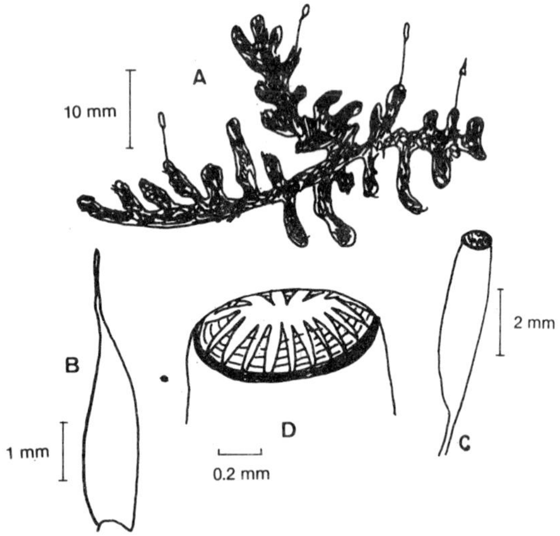

Fig. 24. HAIRY SHORT-BRANCHED WHITE MOSS (*Cladopodanthus pilifer*) A — whole plant; B — leaf; C — capsule; D — peristome.

very easily broken off. Since most of the leaves have broken-off tips, it is almost impossible to find a complete leaf for examination. The stem is triangular in section and bears three rows of leaves; each side corresponding to one row of leaves. The rows are not strictly vertical, except in very young plants. Each row of leaves describes a loose spiral round the stem. The bases of the leaves overlap with mathematical precision, so that there is a zig-zag pattern running down the stem. Each leaf is very thick at the base and bears several enlarged yellow cells on either side. It is sheathing below, gradually narrowing to an upper rigid part. The apex, when not broken off, is acute, entire, or slightly irregular. The nerve occupies two-thirds of the base. The lamina is broader at one side than the other, but has entirely disappeared at the tip. A cross-section of the leaf indicates that it is made of several layers of dead *leucocysts* in which are buried three layers of small, living, green *chlorocysts.* In larger tufts stems may branch and all branches tend to form a smooth cushion. This moss is readily recognized by the three rows of leaves.

The seta may be terminal in position, or may arise laterally between two branches. The capsule is rather slender and upright. The peristome is single, consisting of 16 very short, stumpy, deeply inserted, papillose teeth arranged in pairs. They are not dicranoid. The operculum is pointed, sometimes with a bent tip. The calyptra is hood-shaped and split down one side.

This is one of the easiest of all local mosses to identify. It is readily distinguished from all others by its whitish bluish colour, brittle nature, and three rows of leaves.

Localities: it is widespread on trees in shady areas in both Malaysia and Singapore. It is distributed throughout Southeast Asia, in the Pacific Islands, and in Australia.

POINTED SPLIT-CAP WHITE MOSS

Schistomitrium apiculatum (Fig. 23)

The Pointed Split-Cap White Moss is a small species, forming loose whitish tufts about 2 cm high. The stems have true branching with crowded leaves turned to one side (*falcate*). Leaves are about 2 × 0.6 mm, somewhat concave with a wide ovate base and a flat bent-over tip when dry. Most of the leaf consists of two layers of dead *leucocysts* with a single layer of *chlorocysts.* The lamina is exceedingly narrow, being only 3—4 rows wide at the base and disappearing above.

The seta is rather short, about 3 mm high, smooth or slightly roughened. The capsule is upright and cylindrical. The operculum is a drawn-out conical shape. The calyptra is very characteristic of this type of moss. It forms a complete conical covering, with the base fringed with many hair-like projections (*mitriform*). Its colour is brownish to golden.

Localities: This species is found on the ground and tree trunks above 1,000 m, for example, on Gunong Tahan, the mountains of Pahang and Perak, and Fraser's Hill. It has, however, also been recorded from Johore Bahru but not from Singapore. It is also found in Indonesia and the Philippines.

HAIRY SHORT-BRANCHED WHITE MOSS
Cladopodanthus pilifer (Fig. 24)

The Hairy Short-Branched White Moss is a robust, golden-green plant with upright or creeping branches which bear short shoots covered with very dense leaves. The leaves are rounded oblong to spoon-shaped. The apex is extremely concave and it bears a long hair at the tip equal to one-quarter or one-third of the total leaf length. The nerve is wide, occupying most of the leaf which consists mainly of dead *leucocysts.*

The seta is often borne terminally on short branches, and one plant may bear several capsules. Each seta is about 1 cm high and a little papillose. The capsules are upright and long cylindrical, brownish red, and shiny. The peristome teeth are deeply inserted; some being split half-way and dicranoid.

Localities: Fraser's Hill. It is widely distributed in the mountains of Indonesia.

BLUME'S OUTER NET WHITE MOSS
Exodictyon blumii (Fig. 25)

The Blume's Outer Net White Moss is a slender plant with widely spreading leaves which are laxly placed on the stem. Each leaf is about 4 × 0.5 mm and folded along its entire length when dry. Most of the leaf consists of dead cells, but green *chlorocysts* are exposed on the surface in the region of the nerve, so that this plant appears a little less white than other members of the family. The nerve occupies one-third of the base but extends beyond the lamina. It is spinulose on both sides of the leaf. The capsule has not been seen locally.

Localities: Penang and Tapah; also in the Philippines and Indonesia.

SULLIVANT'S OUTER NET WHITE MOSS
Exodictyon sullivantii (Fig. 26)

The Sullivant's Outer Net White Moss has shorter leaves than the previous one, only 2 mm long, with a lamina extending nearly to the

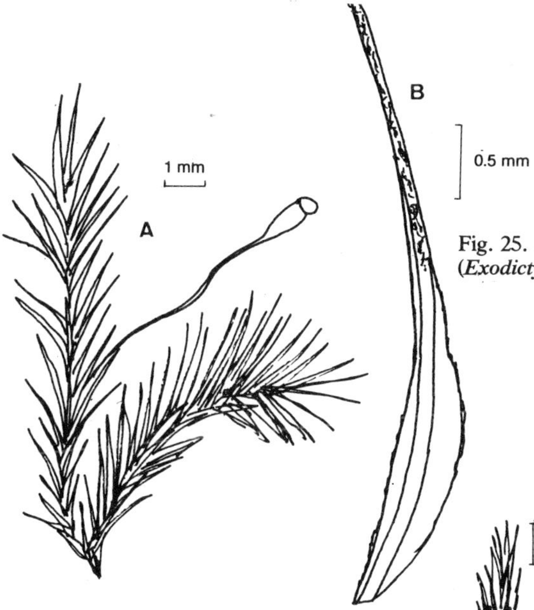

Fig. 25. BLUME'S OUTER NET WHITE MOSS (*Exodictyon blumii*) A — whole plant; B — leaf.

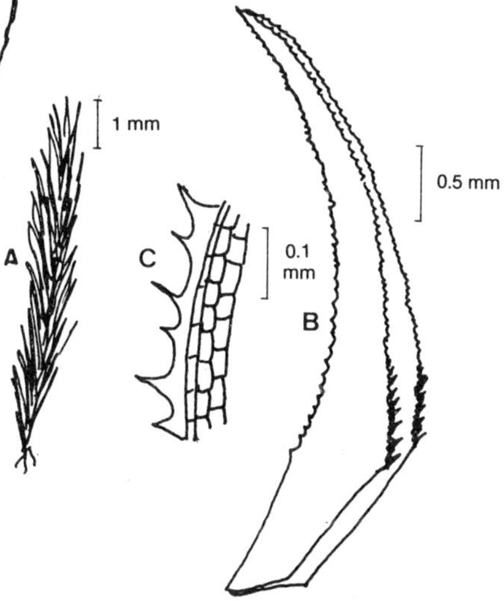

Fig. 26. SULLIVANT'S OUTER NET WHITE MOSS (*Exodictyon sullivantii*) A — whole plant; B — leaf; C — leaf border.

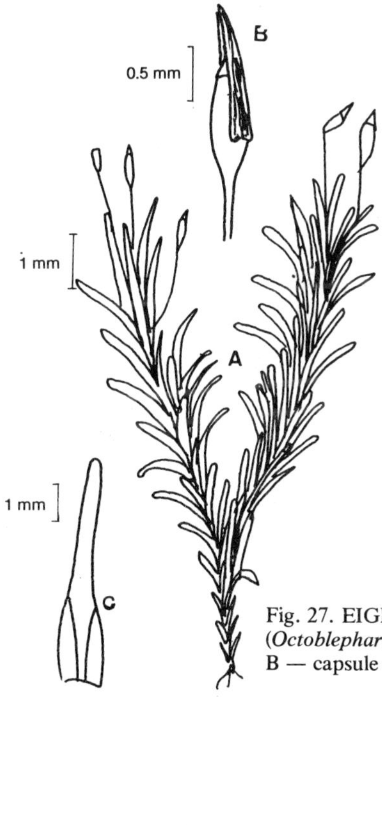

Fig. 27. EIGHT-TOOTHED WHITE MOSS (*Octoblepharum albidum*) A — whole plant; B — capsule with calyptra; C — leaf.

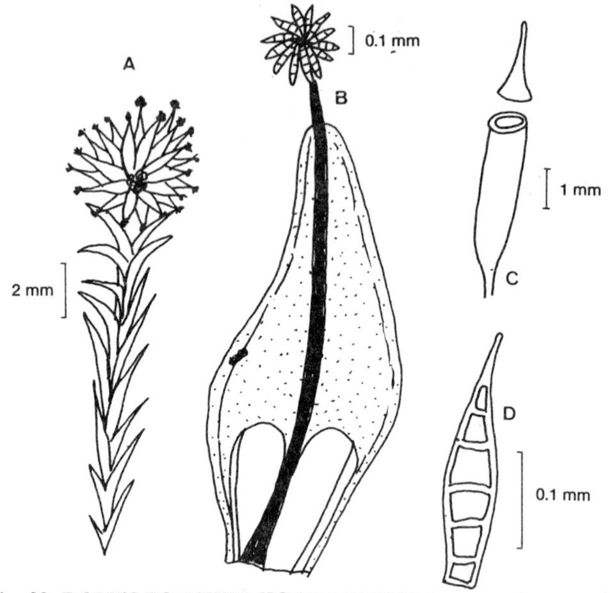

Fig. 28. DOZY'S POUCHED HOOD MOSS (*Calymperes dozyanum*) A — whole plant with gemmae; B — leaf with excurrent nerve and gemmae; C — capsule; D — gemma.

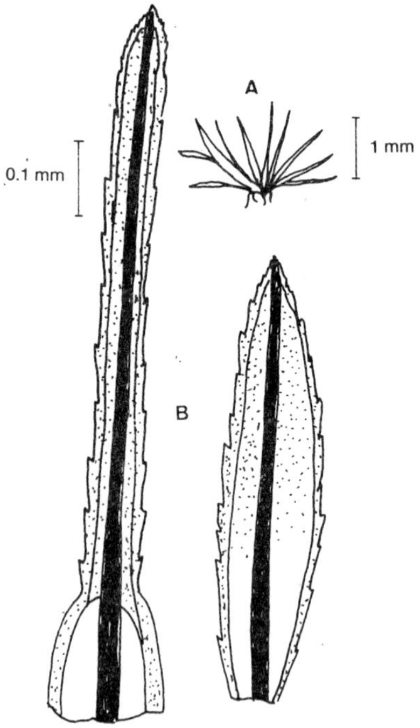

Fig. 29. TOOTHED POUCHED HOOD MOSS (*Calymperes serratum*) A — whole plant; B — leaves.

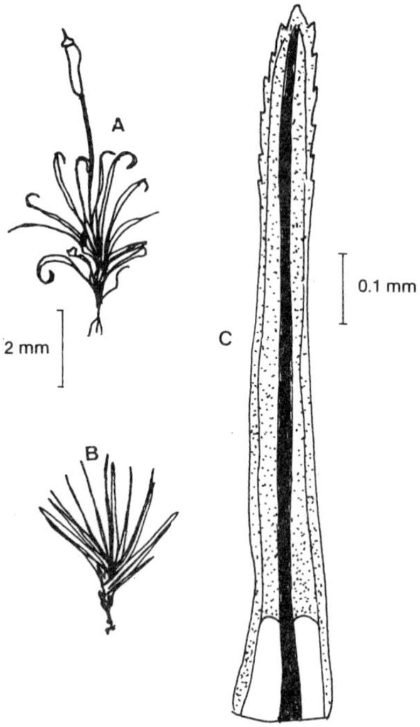

Fig. 30. LONG-LEAVED POUCHED HOOD MOSS (*Calymperes longifolium*) A — whole plant (dry); B — whole plant (wet); C — leaf.

apex. Its leaf bases are somewhat white, and it could be mistaken for a *Syrrhopodon* on first examination; but here the nerve is extensive and the lamina not green.

Localities: This has only been found on Gunong Semanggol, Perak. It is also found in the Philippines and Java.

EIGHT-TOOTHED WHITE MOSS

Octoblepharum albidum (Fig. 27)

The Eight-Toothed White Moss is a sturdy plant forming thick, waxy, greenish white to reddish white cushions. The thick leaves are very rigid and slightly iridescent when dry. Like *Arthrocormus,* they scarcely appear to be real plants at all. The leaves here are however a rounded tongue-shaped, or may be widened at the base. The nerve is broad and thick, occupying about half of the base and three-quarters the upper part. It consists of several layers of dead *leucocysts,* with a single layer of *chlorocysts* sandwiched between them.

The seta is 2—3 mm high. The capsule is erect and oblong. The peristome consists of eight well-spaced peristome teeth. The operculum is short-tipped.

Localities: Widespread as tufts on soil and trees, especially in the lowland of North West Malaysia, but it has also been found on Penang Hill. It is very common on the east coast of West Malaysia, but is apparently absent from the south and from Singapore.

◆ ◆ ◆

POUCHED HOOD MOSS FAMILY

(Calymperaceae)

The Pouched Hood Moss Family is a family of tropical mosses which are found mainly on the trunks and branches of trees. The leaves have a hyaline sheath of colourless rectangular cells. The upper part of the leaf consists of small green lamina cells which extend down on either side of the sheath. The nerve is very prominent and is usually excurrent in the upper leaves, often bearing a cluster of gemmae. The border may have a line of special cells (*teniolae*) just in from the margin.

Most species of this family do not appear to form capsules very often, and the main type of reproduction is by gemmae. When a capsule is produced it is terminal in position with a short upright seta. The calyptra is large, has long folds, and completely covers the capsule. It is persistent and remains in position until the capsule is ripe, when the spores are released by longitudinal splitting of the calyptra and operculum. There are no peristome teeth, but the central mass of tissue (*columella*) is raised up to block the mouth.

*DOZY'S POUCHED HOOD MOSS

Calymperes dozyanum (Fig. 28)

The Dozy's Pouched Hood Moss is one of the commonest of local mosses occurring in open and shady situations on tree bark. It is dull dark green in colour. The stem is upright and may be single or several times branched. It grows to about 1.5 cm high. The leaves are curled up when dry but form a spreading rosette towards the top of the stem when damp. The lower leaves are lanceolate while the upper ones have a long excurrent nerve which ends in a flattened spoon-shaped structure bearing a head of numerous gemmae. The inner basal leaf cells are hyaline and rectangular, being sharply differentiated from the small, round, green cells which make up the rest of the lamina. The hyaline part reaches about one-third up the leaf. There is no border and the margin is quite entire. The upper green laminal cells are very small, rounded or hexagonal, covered with papillae, and packed with chloroplasts. The lower laminal cells which border the sheath may be a little elongated and poor in chloroplasts.

This species is rarely fertile and seems to reproduce mainly by gemmae. The seta is usually terminal on the main stem. The calyptra is bell-shaped, completely covering the capsule. The operculum is short-pointed. The capsule is long, cylindrical, and has no peristome teeth.

Localities: Common on trees in both Singapore and Malaysia. Found also in Sri Lanka and Indonesia.

*TOOTHED POUCHED HOOD MOSS

Calymperes serratum (Fig. 29)

The Toothed Pouched Hood Moss has a very small stem bearing about twenty erect spreading leaves. These are about 1.3 × 1 mm, with a small sheath and a wide border continuing from base to apex. The lamina cells are very small and round in shape. The tip is rather broad, with an excurrent nerve covered with close teeth. These teeth continue from the base to the apex. Young upper leaves may be very small and ovate-lanceolate in shape, while abnormal leaves may have no hyaline sheath but have a gradual transition from colourless cells below to green laminal cells above.

Localities: This species has only been recorded from Singapore and Pahang but may also occur elsewhere in Malaysia. It is also found in other parts of Central and Southeast Asia and in the Pacific Islands.

*LONG-LEAVED POUCHED HOOD MOSS

Calymperes longifolium (Fig. 30)

The Long-Leaved Pouched Hood Moss is a lowly plant with a short stem only 1—4 mm high, which bears long strap-shaped leaves. These leaves are about 12 × 0.4 mm, and curved or curled when dry. The sheath is exceedingly small, about 0.25 × 0.1 mm. The lower three-quarters of the margin is entire but the upper one-quarter is toothed. There is a distinct border stopping short of the tip. The nerve is not normally excurrent.

Localities: This probably occurs throughout Malaysia and Singapore on trees. It is recorded from Singapore, Pahang, Penang, and Kedah.

*EXTENDED POUCHED HOOD MOSS

Calymperes porrectum (Fig. 31)

The Extended Pouched Hood Moss is an upright plant with a stem about 1.5 cm high. The leaves consist of an ovate-lanceolate base which is scarcely separated by a "shoulder" from the upper lamina. The leaf is mostly narrowed evenly from base to tip, and is about 7 × 2 mm. Some leaves are completely entire, but normally there are teeth on the border. The apex is acute and pointed. The nerve is prominent, papillose behind, and not normally excurrent. The sheath is rather small, extending to one-third the length of the leaf. It consists of hyaline, rectangular cells. In young leaves, *teniolae* continue uninterrupted from the base to the tip, and they delimit the border which consists of hyaline cells plus the vermiform teniolae. In older leaves, the teniolae may break above the shoulder so that the border becomes transformed into green rectangular cells and disappears before the tip. In other cases it may be retained and the teniolae become indistinct.

Localities: On trees in Singapore and Penang. It is also found in Indonesia and the Pacific Islands.

*CURLY POUCHED HOOD MOSS

Calymperes recurvifolium (Fig. 32)

The Curly Pouched Hood Moss is a tall plant, 2—5 cm high, with very curly, squarrose leaves when dry. The leaves are ovate-lanceolate, about 7 × 1 mm. There is no distinct shoulder, and the sheath occupies about one-third the length of the leaf. The nerve is extremely papillose, excurrent, and toothed; or bearing gemmae. The sheath is indistinctly separated from the lamina above. Elongated teniolae continue from base to tip. The sides of the sheath are bordered with lamina cells, elongated hyaline cells, and teniolae; but the elongated hyaline cells disappear half-way up the sheath.

Localities: Singapore; Gunong Pulai, Johore; Waterfall Gardens, Penang. Also found in Indonesia and Pacific Islands.

*TAHITAN POUCHED HOOD MOSS

Calymperes tahitense (Fig. 33)

The Tahitan Pouched Hood Moss consists of upright plants about 1.2 cm tall, usually with some true branches present. The leaves are 7 × 2 mm, strap-shaped from an oval base. The basal part is almost entirely occupied by a hyaline sheath consisting of large rectangular hyaline cells, outside which is a prominent thickened border about four cells wide, and this extends from the base to the tip to the leaf. The upper part of the leaf is about 5 × 0.75 mm and consists of small thick-walled green cells which are covered with small papillae. The border is distinct throughout and bears a few teeth. The tip is wide and rounded with the nerve extending to the tip.

Localities: This species has only been recorded from Singapore and Perak.

*HAMPE'S POUCHED HOOD MOSS

Calymperes hampei (Fig. 34)

The Hampe's Pouched Hood Moss is a tall plant about 2 cm high, bearing broad lanceolate leaves in which the sheath is very prominent. The upper part of the leaf is often rolled into a canal-shaped structure. The sheath occupies about one-third of the leaf and consists of hyaline cells. The margin bears many small teeth. The nerve is not normally excurrent and also bears small teeth. There is a prominent border with a row of teniolae stretching from base to tip.

Localities: This species is recorded from Singapore, Johore, Penang, and Kedah.

*NICOBAR POUCHED HOOD MOSS

Calymperes nicobarense (Fig. 35)

The Nicobar Pouched Hood Moss is an upright plant about 1.5 cm tall. It has lanceolate leaves with the sheath occupying about one-quarter the distance from the base to the apex. There is a strong border of rect-angular cells with vermiform teniolae extending three-quarters along the

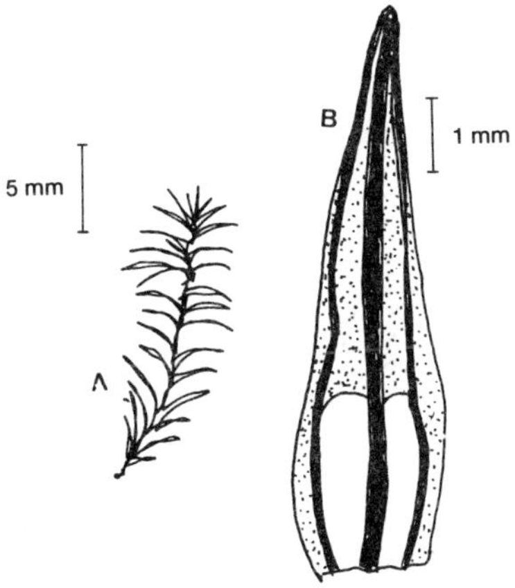

Fig. 31. EXTENDED POUCHED HOOD MOSS (*Calymperes porrectum*) A — whole plant; B — leaf.

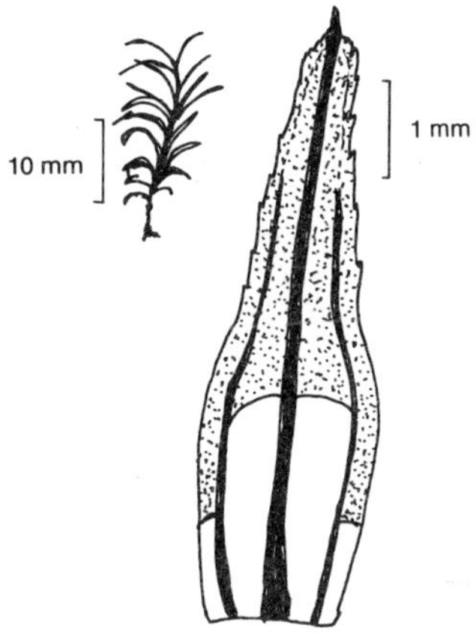

Fig. 32. CURLY POUCHED HOOD MOSS (*Calymperes recurvifolium*) A—whole plant; B—leaf.

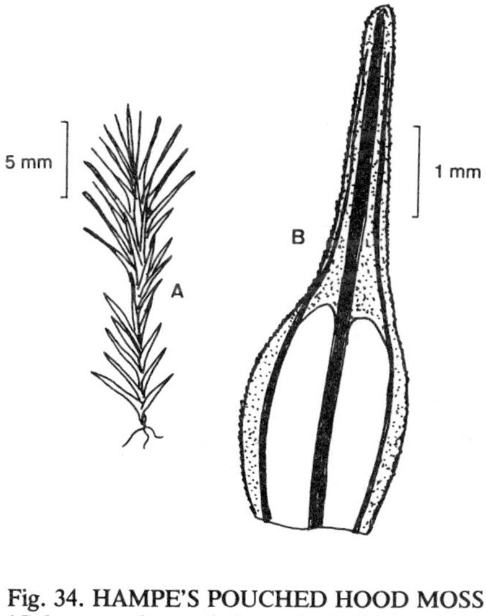

Fig. 34. HAMPE'S POUCHED HOOD MOSS (*Calymperes hampei*) A — whole plant; B — leaf.

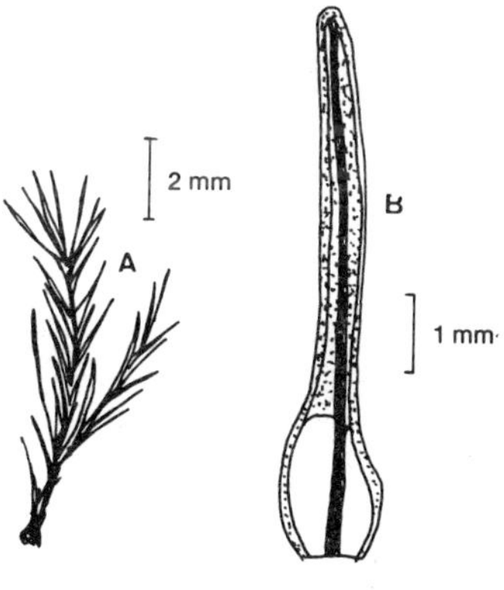

Fig. 33. TAHITAN POUCHED HOOD MOSS (*Calymperes tahitense*) A — whole plant; B — leaf.

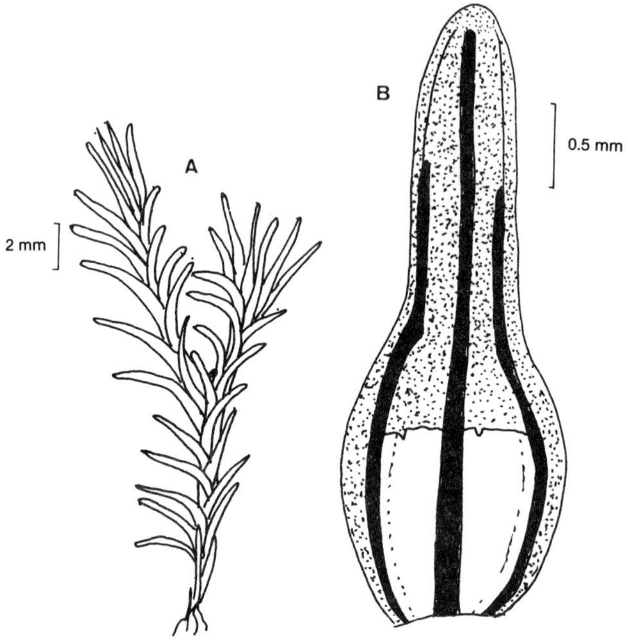

Fig. 35. NICOBAR POUCHED HOOD MOSS (*Calymperes nicobarense*) A — whole plant; B — leaf.

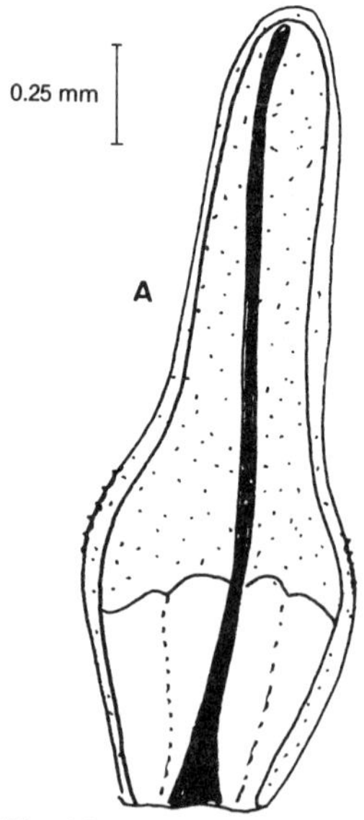

Fig. 36. DELICATE POUCHED HOOD MOSS (*Calymperes tenerum*) A — leaf.

Fig. 37. TJIBODAS POUCHED HOOD MOSS (*Calymperiopsis tjibodensis*) A — whole plant; B — leaf; C — gemmae.

outside of the border. The lamina cells are green and hexagonal, while the sheath is made of small hyaline cells at the base which become larger above. The border ceases just before the tip, but the mid-rib is usually excurrent.

Localities: Numerous records from Singapore. Also found in Pulau Langkawi.

*DELICATE POUCHED HOOD MOSS

Calymperes tenerum (Fig. 36)

The Delicate Pouched Hood Moss is a small plant about 1 cm high, forming soft mats or cushions. The leaves are about 2 mm long, with an ovate base narrowing to a wide tongue-shaped apex. There is a distinct border. The sheath occupies one-third of the leaf and is divided into the inner sheath of large rectangular cells, and the outer sheath of smaller cells. The lamina cells are small and papillose. The margin is toothed.

Localities: Singapore. Also found in the Philippines, India, and the Pacific Islands

*TJIBODAS POUCHED HOOD MOSS

Calymperiopsis tjibodensis (Fig. 37)

The Tjibodas Pouched Hood Moss occurs in thick, low, bright green tufts. The stems are usually branched but sometimes single, about 1 cm high, round in section, and thickly leaved above. The leaves are narrow lanceolate, sharply tipped, about 2.5 × 0.5 mm, and somewhat keeled. The hyaline sheath occupies from one-quarter to one-third the total length of the leaf, and is made of colourless rhombic cells. The lamina cells are intensely green, four-sided, and bear large mamillate papillae. The mid-rib is extremely prominent and ends in a short thorn-tip. A cluster of gemmae occurs in the upper leaves half-way along the ventral side, above the sheath. There is a prominent entire border.

Localities: On trees in Singapore. Originally found in Java.

HORIZONTAL TOOTH MOSS FAMILY

(Syrrhopodontaceae)

The Horizontal Tooth Moss Family, which only occurs in the tropical and sub-tropical regions of the world, has a large number of species in this area which are very difficult to distinguish from one another. Since the taxonomy has not been fully worked out, only an imperfect treatment can be given here. Horizontal Tooth Mosses mostly occur as tufts or cushions on bark or, rather more rarely, on the ground. The stem is single or has true branching. It is usually upright, except for *Thyridium* where it creeps along the ground. The leaves are very similar in structure to those of the Calymperaceae, and have a short or long hyaline sheath at the base. The lamina usually bears a border of thickened marginal cells which may bear short or long thorns or, occasionally, long cilia or teeth, rarely entire.

The capsule is usually borne terminally on a long seta. It is upright, ovoid-cylindrical, and without a distinct annulus. The peristome is always present, and this distinguishes it clearly from the previously family. It consists of 16 papillose, deeply inserted peristome teeth, with prominent transverse bars. The operculum, which is much shorter than the capsule in length, has an abrupt beak. The calyptra is smooth and hood-shaped. It splits up one side and is lost before maturity. This is another difference from the Calymperaceae.

*ROLLED HORIZONTAL TOOTH MOSS

Syrrhopodon revolutus (Fig. 38)

The Rolled Horizontal Tooth Moss is a glossy, fragile, pale coloured plant growing in compact cushions on the bark of trees. The stem is 3 cm high and several times branched. It may lose most of its leaves below. The leaves are closely overlapping and less than 2 mm long. Their shape is somewhat variable, but they have a widened lower portion with a prominent hyaline sheath which may reach up to two-thirds the entire length. The nerve is prominent but slender, and ends just below the apex. There are green cells on either side of the sheath, while those of the upper lamina are rounded and green.

The seta is smooth and slender, about 5 mm long. The capsule is urn-shaped and somewhat asymmetrical. The operculum has a long pointed tip. Half of the capsule is hidden in a densely hairy calyptra which has a fringed base and one split extending up the side. The 16 peristome teeth are undivided, yellow, and papillose.

Localities: Recorded from Singapore, Taiping, and Penang.

GARDNER'S MOSS

Syrrhopodon gardneri (Fig. 39)

The Gardner's Moss is an upright moss forming thick, greenish-brown tufts on tree trunks. When dry the leaves are twisted and curled up, but when damp they are fully expanded, forming a rosette at the top of the stem..Individual plants are about 2 cm high, rarely branched, and sometimes bent upwards from a prostrate base. The leaves are long lanceolate, with a rounded tip. They are scarcely narrowed at the base. The sheath occupies about one-fifth to one-quarter of the total length of the leaf. The rib is very prominent and can even be seen with the naked eye in wet specimens, since the drak rib contrasts with the translucent colour of the lamina. The leaf has a distinct border which, in the sheath part of the leaf, consists of long thick-walled cells which merge into small cubical cells above. The cubical cells only differ from ordinary lamina cells by the thickness of the wall and the intensity of the green colour. This border is technically intra-laminal, that is, it does not form the outermost layer of the leaf since there is yet another layer of thin-walled marginal cells. The colourless cells making up the sheath are very large and prominent. The margin of the leaf is distinctly toothed, especially near the leaf tip.

Localities: Kuala Tahan. Widely distributed in Indonesia, Sri Lanka and Nepal.

*REDDISH HORIZONTAL TOOTH MOSS

Syrrhopodon rufescens (Fig. 40)

The Reddish Horizontal Tooth Moss occurs as thick, glossy, greenish brown cushions which often have an inner reddish or golden tinge. The upright stems are sparingly dichotomously branched, and about 2 cm high. The leaves are erect-spreading, 1.5 mm long, narrow lanceolate, and with even flexuose tips. Both the border and mid-rib are prominent. Green laminal cells are restricted to the upper one-third of the leaf, the lower two-thirds being made of colourless hyalocysts. The capsule is borne on a thin, golden yellow seta which is about 8 mm tall. It is erect and elongated under the mouth. The calyptra is papillose and entire when young, but splits vertically as it matures. The peristome consists of 16 lanceolate teeth.

Localities: On trees at 100 m elevation on Bukit Timah Hill, Singapore; Botanic Gardens, Singapore. Also found in the Philippines, Java, and the Marianne islands.

*WHITE-SHEATHED HORIZONTAL TOOTH MOSS

Syrrhopodon albo-vaginatus

The White-Sheathed Horizontal Tooth Moss is found in rather sparse tufts of a dull brownish green colour. The stem is about 1.5 cm high, and may be branched. The leaves are tongue-shaped with an erect, white, sheathing base and a blunt apex. Each leaf is 4 mm long, densely papillose, and with a coarsely serrate border above. The prominent mid-rib is serrate behind. The seta is slender, erect, 5 mm tall, and bears an erect, glossy capsule.

Localities: Recorded from Singapore, Kuala Lipis, and Sungei Siput. Probably widespread in Malaysia. Found also in the Philippines, Indonesia, the Moluccas, and Fiji.

*HAIRY HORIZONTAL TOOTH MOSS

Syrrhopodon ciliatus (Fig. 41)

The Hairy Horizontal Tooth Moss occurs in dirty green tufts with visible whitish leaf sheaths. The stem is 1—2 cm high, single or branched, with leaves very crowded at the top, slightly concave, and forming a rosette. It differs from all other species of Horizontal Tooth Moss by the long hairs borne on the hyaline border. However, in Singapore, an unusual variety occurs (var. *pseudopodianus*) where the hairs are replaced by long ten-celled gemmae. The sheath occupies half to one-third of the length of the leaf. The lamina cells are small, smooth, rounded, and four-sided. The seta is red, 5 mm long, and bears a long, ovoid capsule at the top. The peristome consists of 16 deeply inserted teeth.

Localities: Botanic Gardens, Singapore; Kuantan; Sungei Krian Estate. Also found in the Philippines and Indonesia.

*SPINY HORIZONTAL TOOTH MOSS

Syrrhopodon spiculosus (Fig. 42)

The Spiny Horizontal Tooth Moss is a slender plant similar to the Hairy Horizontal Tooth Moss, but the stems are taller, up to 3 cm high, yellowish green above and brownish below. The leaves are narrower and they have a stunted tip. The hyaline border is rather narrow and extends only three-quarters of the way up the leaf. A few spine-like teeth are borne on the upper part of the sheath. The mid-rib is very prominent.

Localities: Kranji and Bukit Timah, Singapore; Pulau Tioman; Pulau Rumbia; Kedah Peak. Also found in Indonesia and the Philippines.

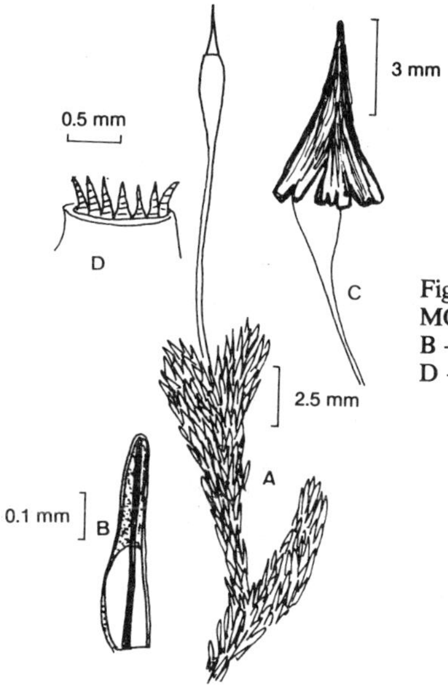

Fig. 38. ROLLED HORIZONTAL TOOTH MOSS (*Syrrhopodon revolutus*) A — whole plant; B — leaf; C — capsule with calyptra; D — peristome teeth.

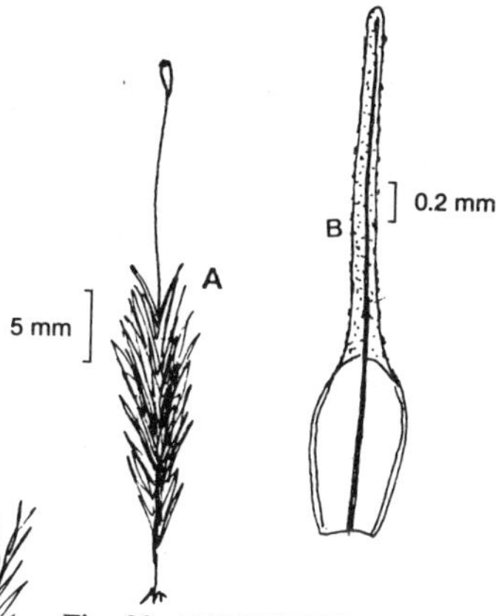

Fig. 39. GARDNER'S MOSS (*Syrrhopodon gardneri*) A — whole plant; B — leaf.

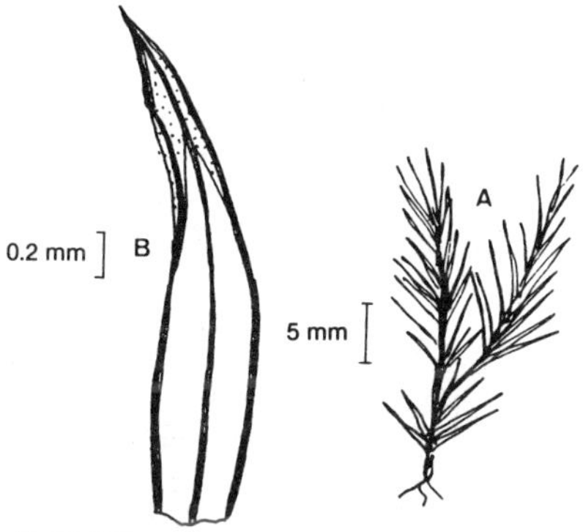

Fig. 40. REDDISH HORIZONTAL TOOTH MOSS (*Syrrhopodon rufescens*) A — whole plant; B — leaf.

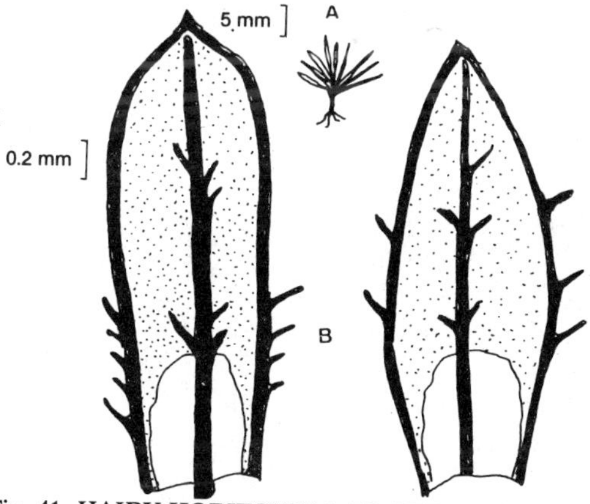

Fig. 41. HAIRY HORIZONTAL TOOTH MOSS (*Syrrhopodon ciliatus*) A—whole plant; B—leaves.

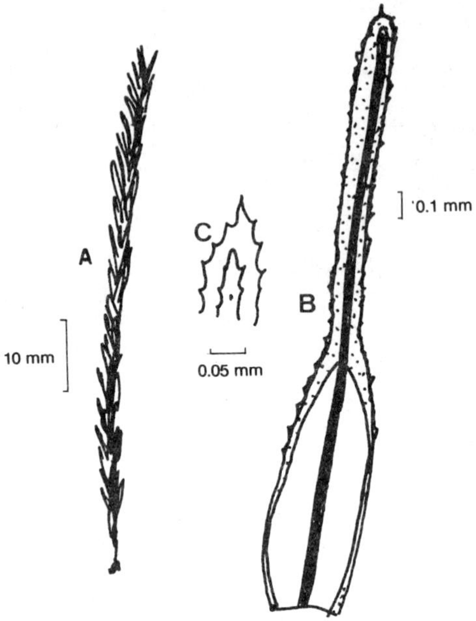

Fig. 42. SPINY HORIZONTAL TOOTH MOSS (*Syrrhopodon spiculosus*) A—whole plant; B—leaf; C—leaf tip.

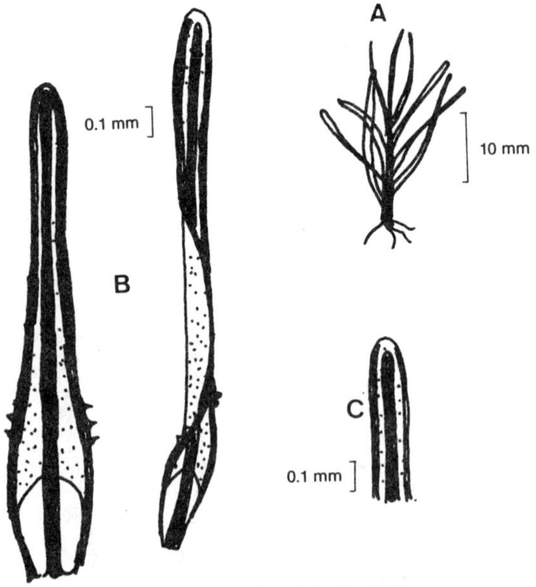

Fig. 43. YELLOW HORIZONTAL TOOTH MOSS (*Syrrhopodon croceus*) A — whole plant; B — leaves; C — leaf tip.

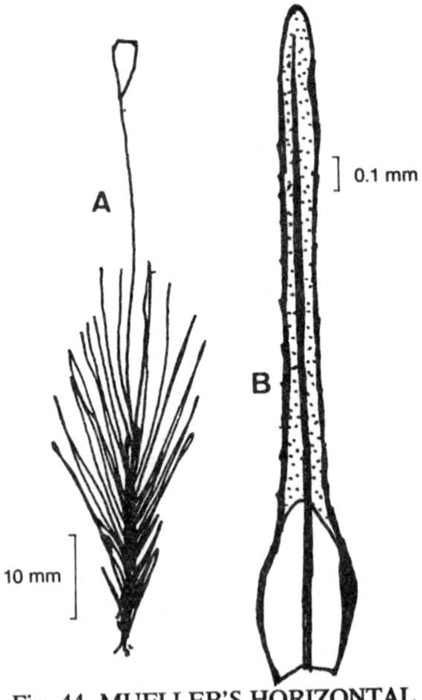

Fig. 44. MUELLER'S HORIZONTAL TOOTH MOSS (*Syrrhopodon muelleri*) A — whole plant; B — leaf.

*YELLOW HORIZONTAL TOOTH MOSS

Syrrhopodon croceus (Fig. 43)

The Yellow Horizontal Tooth Moss forms rather loose tufts of a dull reddish yellow colour. The stems are up to 3 cm high, single or once-branched. The leaves are about 6 mm long, rather stiff, erect-spreading or slightly curved. The base is expanded and orange red in colour, but the upper part of the leaf is linear and forms a narrow channel. The margin bears short paired teeth with a special group of teeth located at the top of the sheath. The sheath occupies about one-third the length of the leaf. The lamina cells are small, rounded, and papillose. The border extends from base to tip. The mid-rib is prominent and orange red in colour.

Localities: Singapore; Gunong Belumut; Gunong Panti; Gunong Pulau; Gunong Tahan; Sungei Ketah; Penang. Also found in Sri Lanka, Indonesia, and Oceania.

*MUELLER'S HORIZONTAL TOOTH MOSS

Syrrhopodon muelleri (Fig. 44)

The Mueller's Horizontal Tooth Moss is an upright moss forming loose, brownish green tufts, 1—2.5 cm high. The plants have a bushy appearance with crowded erect-spreading leaves above. Each leaf is usually about 5—7 mm long but is sometimes even longer. The base of the leaf is ovate, but this narrows into a linear upper lamina. The margins are entire except for a few teeth at the tip. The mid-rib is smooth and prominent. The lamina cells are irregular, rounded, and more or less four-sided. The thickened border extends almost to the tip. The seta is scarely 5 mm long, reddish brown in colour, and bearing an ovoid capsule which is about 1 mm long. The peristome consists of a row of narrow, lanceolate, papillose teeth.

Localities: On tree trunks in Singapore, Gunong Pulai, Temengoh, Gunong Batu Puteh, Penang Hill, Gunong Tahan. Also found in Sri Lanka, Thailand, Indonesia, the Philippines, Australia, and Oceania.

*RIDLEY'S HORIZONTAL TOOTH MOSS

Syrrhopodon ridleyi

The Ridley's Horizontal Tooth Moss is very similar to Reddish Horizontal Tooth Moss but the leaf tips are less flexuose, although they are defintely not adpressed. This has not been found recently in Singapore or Malaysia, and may not be a distinct species.

Localities: Bukit Timah, Singapore, collected by H.N. Ridley in 1908; Pulau Penang at 800 m in 1915.

*DENSE HORIZONTAL TOOTH MOSS

Syrrhopodon confertus

The Dense Horizontal Tooth Moss may not be a distinct species since it only differs from other Horizontal Tooth Mosses in its relatively large size and very thickly inserted leaves. Its hyaline sheath extends three-quarters the leaf length.

Localities: Singapore; Gunong Tahan; Fraser's Hill; Gunong Angri, Negri Sembilan.

*YELLOW GATE MOSS

Thyridium flavum (Fig. 45)

The Yellow Gate Moss species grows in dense mats of loose green plants. The primary stem, which is creeping, is about 4 cm long. It bears numerous rhizoids and upright secondary stems, 1 cm tall, each bearing a crowded rosette of leaves. The leaves have a short clasping base and a lanceolate upper lamina which is sharply tapering. The total length of the leaf is 1.25 mm. The border is prominent, toothed, golden yellow in colour, and extends three-quarter way up the leaf. It is strongly undulate and bears small teeth. The sheathing base consists of thin-walled colourless cells with small green laminal cells above. The seta is red, 4–5 mm tall, and bears a long cylindrical capsule. The peristome teeth are deeply inserted and consist of a single row of 16 lanceolate, papillose teeth.

Localities: Singapore. Also found in the Philippines and Java.

*BUNDLED GATE MOSS

Thyridium fasciculatum (Fig. 46)

The Bundled Gate Moss has branches arising in bundles (*fascicles*), forming small clusters about 2 cm long on the upper side of the horizontal creeping stem. The leaves are about 3–5 mm long and 1.5 mm broad, with a wide sheathing base slowly tapering into a broad lanceolate shape above. The tip is pointed and bears faint teeth. The prominent border is golden in colour but disappears below the tip. The base consists mainly of colourless hyalocysts. The green laminal cells above are small and oval to square in shape. The mid-rib is prominent and just reaches the tip. The seta is red in colour, 4—5 mm high, and bears a long ovoid capsule. The peristome is deeply inserted and consists of 16 lanceolate, papillose teeth.

Localities: On trees, in Singapore, Gunong Pulai, Ginting Simpah, Gunong Tahan, Sungei Bertam. Also found in South India, Sri Lanka, Vietnam, Indonesia, Australia, Chile, and Oceania.

*CARDOT'S GATE MOSS

Thyridium cardotii (Fig. 47)

The Cardot's Gate Moss is a green to golden-coloured moss with a creeping main stem, 3—5 cm long, forming very loose patches. The side branches are short or long, irregular, and thickly covered with leaves which are spreading when wet. They have a broad sheathing base and an upper lanceolate part. There is a prominent hyaline border which is ten cells wide at the base, but is gradually reduced to one cell wide near the tip. The border is serrate from above the sheath to the tip. The basal part consists mainly of hyaline cells. The green laminal cells above are small, ovate to quadrate. The mid-rib is prominent and reaches the tip. Upper leaves frequently bear small, branched gemmae and red hairs on the inner side of the mid-rib just below the tip. The tip is never excurrent (cf *Calymperes*).

Localities: Singapore. Also found in West Java.

*CREEPING GATE MOSS

Thyridium repens

The Creeping Gate Moss is similar to the other species of Gate Moss, but the broad hyaline border, which extends to the tip, bears long hair-like teeth. The lamina cells are oval in shape and thick-walled.

Localities: Botanic Gardens, Chua Chu Kang, Pasir Panjang, Bukit Timah, Bukit Panjang, Kranji, Singapore; Malacca. Also found in Sri Lanka and Banka.

*MAN'S GATE MOSS

Thyridium manii

Man's Gate Moss is similar to Creeping Gate Moss but with a narrow border reaching only two-thirds the length of the leaf. The border is scarcely toothed. The lamina cells are little thickened, irregular, rounded or rarely oval.

Localities: Botanic Gardens, Seletar, Tanjong Katong, Pulau Ubin, Kallang, Singapore; Penang; Gunong Tahan; Maxwell's Hill.

*WAVY GATE MOSS

Thyridium undulatum (Fig. 48)

The Wavy Gate Moss is a robust creeping species, bearing upright branches about 1.5 cm tall. These branches are sometimes sparsely branched. The leaves are very wavy when dry, but spread out when moist. Each leaf has a wide ovate base and extended upper part. The hyaline sheath is less than a quarter the length of the leaf. The rest of the lamina cells are small, green, and papillose. The seta arises terminally on lateral branches, and is about 1.5 cm tall. The capsule is unsymmetrical, urn-shaped, with a long pointed operculum covered with the pointed calyptra split down one side when young. The peristome teeth are 16 crowded triangular structures which have horizontal barred thickening.

Localities: Singapore and Penang. Found also in Java.

SPLIT TOOTH MOSS FAMILY
(Fissidentaceae)

The Split Tooth Moss Family is a cosmopolitan family of mosses, normally found on earth or rocks in open or shady situations. It is an easy family to recognize because it bears its leaves strictly in two rows. In all other local families of mosses, leaves are in more than two rows, or are spirally arranged. Members of the Split Tooth Family have a prominent nerve in their leaves which distinguishes them from leafy liverworts. Each leaf is attached vertically to the stem, and the upper part of the lamina is split into two layers to form separate laminae which clasp the stem. Under the microscope, this little "pocket" being two cells thick is darker than the rest of the lamina which is only one cell thick.

The seta is usually borne terminally on the stem. The capsule has 16 peristome teeth which are split vertically into two halves so that 32 arms are formed. This is why these mosses are known as Split Tooth mosses.

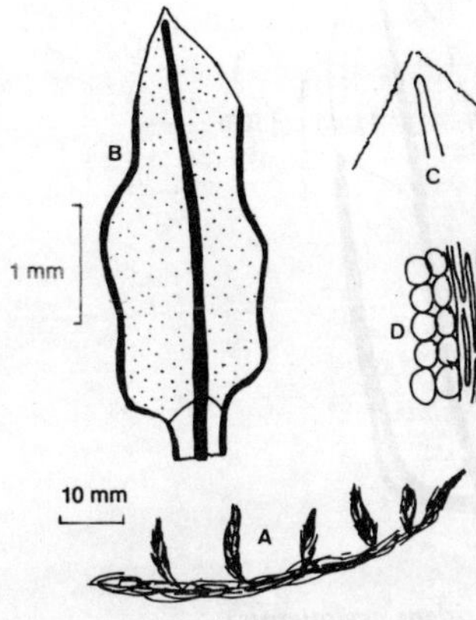

Fig. 45. YELLOW GATE MOSS (*Thyridium flavum*) A — whole plant; B — leaf; C — leaf tip; D — leaf margin.

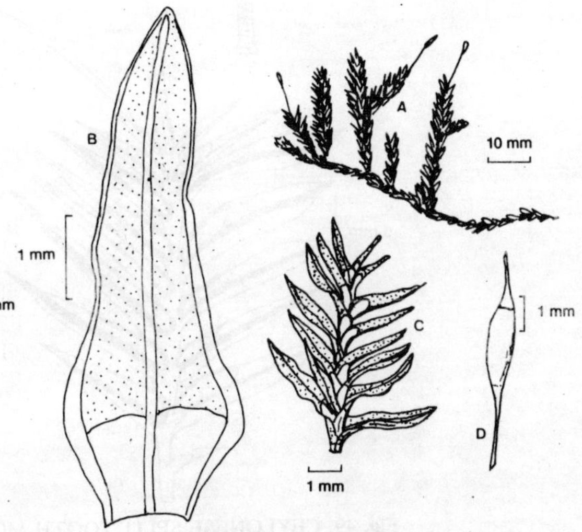

Fig. 46. BUNDLED GATE MOSS (*Thyridium fasiculatum*) A — whole plant; B — leaf; C — part of upright branch; D — capsule.

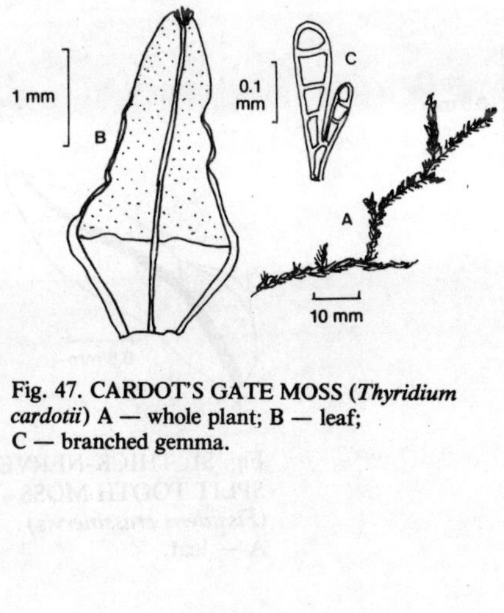

Fig. 47. CARDOT'S GATE MOSS (*Thyridium cardotii*) A — whole plant; B — leaf; C — branched gemma.

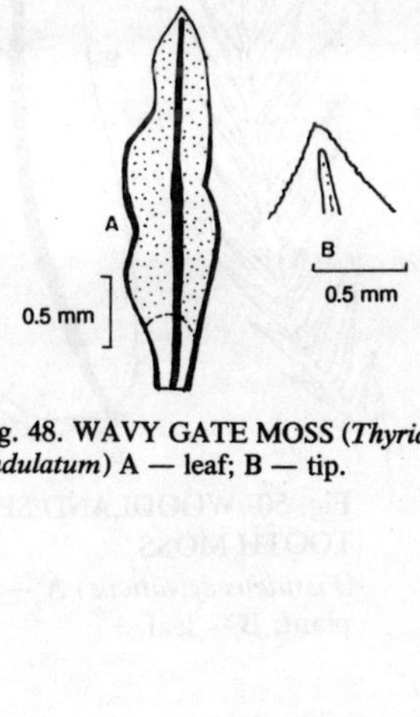

Fig. 48. WAVY GATE MOSS (*Thyridium undulatum*) A — leaf; B — tip.

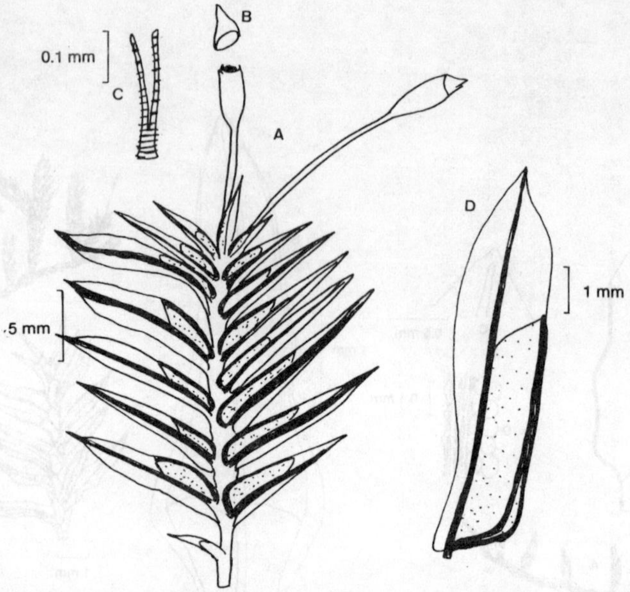

Fig. 49. CEYLONESE SPLIT TOOTH MOSS (*Fissidens ceylonensis*)
A — whole plant; B — operculum; C — peristome tooth; D — leaf.

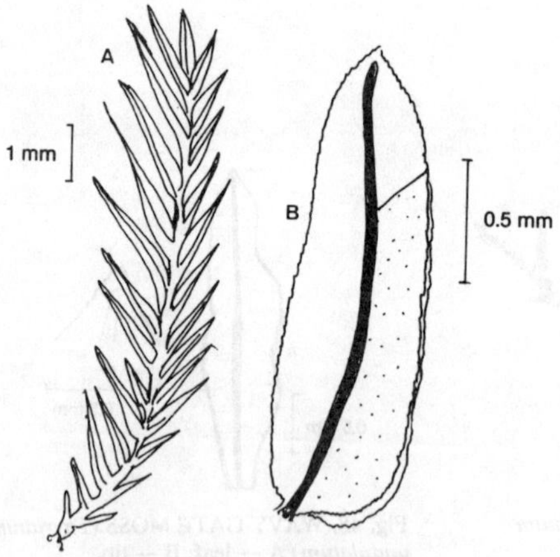

Fig. 50. WOODLAND SPLIT TOOTH MOSS
(*Fissidens sylvaticus*) A — whole plant; B — leaf.

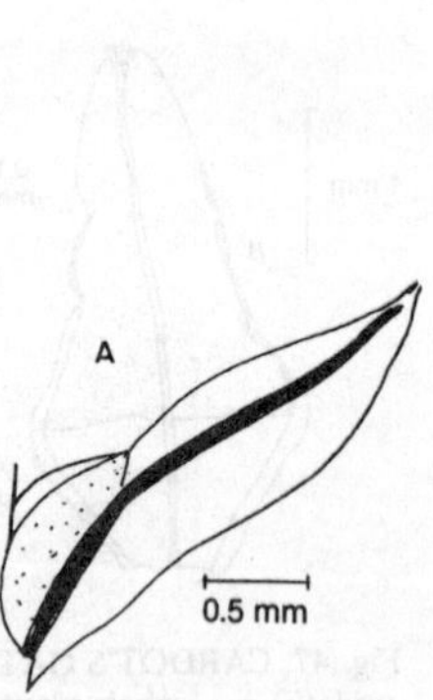

Fig. 51. THICK-NERVED SPLIT TOOTH MOSS
(*Fissidens crassinervis*)
A — leaf.

*CEYLONESE SPLIT TOOTH MOSS

Fissidens ceylonensis (Fig. 49)

The Ceylonese Split Tooth Moss is a very common but minute plant, only 2—6 mm high. It may occur as individual plants or groups of small tufts. It is normally found on bare sloping ground and is dull bluish-green to dark olive green in colour. It differs from other local species in having a prominent white or yellowish nerve. The lower stem may be branched. Each branch has 7—15 pairs of leaves which are strictly arranged in two rows. The lower leaves are reduced and ovate in shape. The upper leaves are larger and longer. The nerve is straight and whitish, in contrast to the dark, opaque, green lamina. There is a short pointed tip. The double portion of the lamina reaches about two-thirds up the upper lamina, and has a whitish border of several rows of hyaline cells.

The seta may be terminal or lateral. It is about 3 mm long. The capsule is urn-shaped and bears a short pointed lid. Each tooth is split into two long arms more than half-way to the base (*dicranoid*).

Localities: Found throughout Malaysia and Singapore on soil and rock in the lowlands. Widely distributed in India, Sri Lanka, Thailand, Laos, Vietnam, Khmer, Indonesia, the Philippines, and New Zealand.

*WOODLAND SPLIT TOOTH MOSS

Fissidens sylvaticus (Fig. 50)

The Woodland Split Tooth Moss is a robust species, from 3 mm to 1.5 cm high, which forms thick tufts of a dirty pale-green colour on earth or rocks. The stems are normally single but occasionally branched into two and bent upwards from the base. Leaves are in 8—12 pairs. Lower leaves are triangular, being progressively smaller towards the base; while upper leaves are twisted to one side, are long triangular-lanceolate, and about 2 × 0.5 mm. The leaves have no border. The double portion of the upper lamina is about two-thirds the total length of the leaf. The lamina cells are opaque, green, and papillose. The margin is crenate, with marginal cells slightly less rich in chloroplasts. The nerve is prominent, opaque, and ends in the short tip.

Localities: Found in Ulu Gombak, Selangor; Kuala Tahan and Sungei Teku, Pahang; Penang; Pulau Dayang Bunting and Kuah, Kedah; Bukit Timah, Singapore. Most of the localities are in forested areas. Distribution is extremely wide, stretching from Central Africa, through the Indian region to Southeast and East Asia. A distinct variety (var. *irroratus*) with longer leaves and a faint border is found in moving water near Klang Gates, Selangor; while var. *fontanus* with much smaller leaves and differing in colour has been found at Chaga Serai, Pahang, submerged in water in the Sungei Tahan.

*THICK-NERVED SPLIT TOOTH MOSS

Fissidens crassinervis (Fig. 51)

The Thick-Nerved Split Tooth Moss is found as small inconspicuous solitary plants forming dense patches on earth. It is pale green in colour and the maximum height of the sterile plant is 4 mm. The leaves are in about 10 pairs. They are ovate-lanceolate and about 1.5 × 0.8 mm. The apex is acute. There is no border. The double portion is only one-third of the upper lamina. The leaf cells are small, transparent, and not papillose. The prominent nerve is red or yellow in colour, and characteristically has a sudden bend half-way along. It ends below the tip. The margin has small teeth. Lower leaves are very much reduced and in extreme cases they consist only of the double portion.

Localities: The species have been found in forested areas in Singapore, Johore, Pahang, Kedah, and Perlis; and in the caves in Selangor and Pahang. It is widely distributed in Southeast Asia. It has frequently been found on dried mangrove mud, including the burrows of the mud lobster, *Thalassina anomala.* Although it may be confused with the next species, *Fissidens mittenii,* it can be distinguished by the bent nerve which ends well below the tip. The leaf cells are also smaller and not differentiated at the base. Although it is a species of moderate shade it can tolerate full sunlight and has a wide distribution in cleared areas in Singapore.

*MITTEN'S SPLIT TOOTH MOSS

Fissidens mittenii (Fig. 52)

The Mitten's Split Tooth Moss forms tufts of many individuals 2–5 mm high, which are reddish green to reddish brown in colour. It is far more gregarious than the other species, and is normally found on earth banks, forming extensive patches. The stem has about 8 pairs of ovate-lanceolate leaves. Upper leaves are 0.75 × 0.2 mm, with the double portion half the length of the upper lamina. No border is present. The lamina cells are rounded, irregular, not papillose, and transparent. The nerve is prominent, yellow, not bent, and extending just beyond the tip to give a short point. The leaf base is very unequal, with the double portion wider. The basal cells are much larger and rectangular.

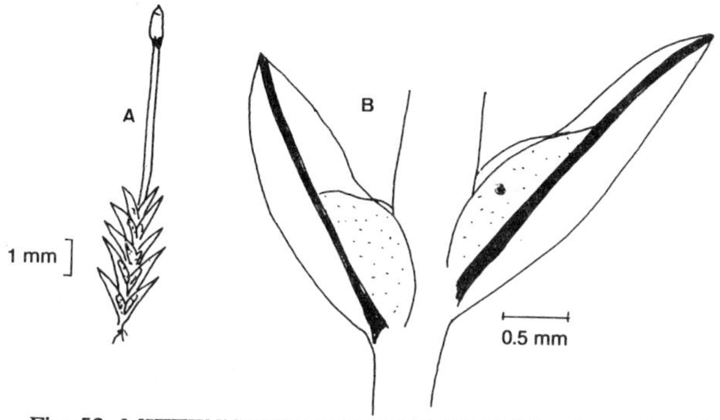

Fig. 52. MITTEN'S SPLIT TOOTH MOSS (*Fissidens mittenii*)
A — whole plant; B — two leaves.

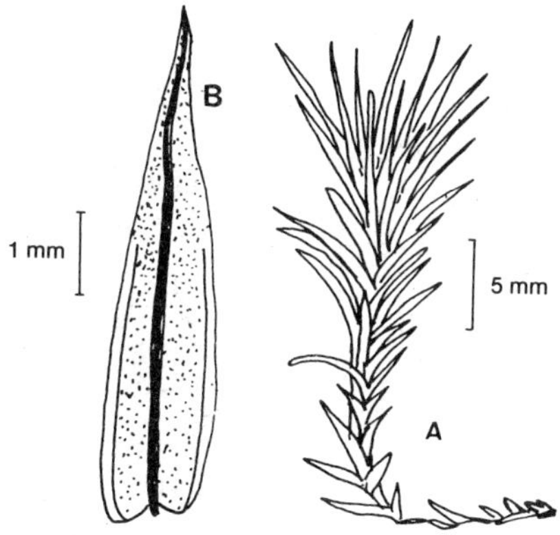

Fig. 53. SNOW-BORDERED POTT MOSS (*Chionoloma latifolium*) A — whole plant; B — leaf.

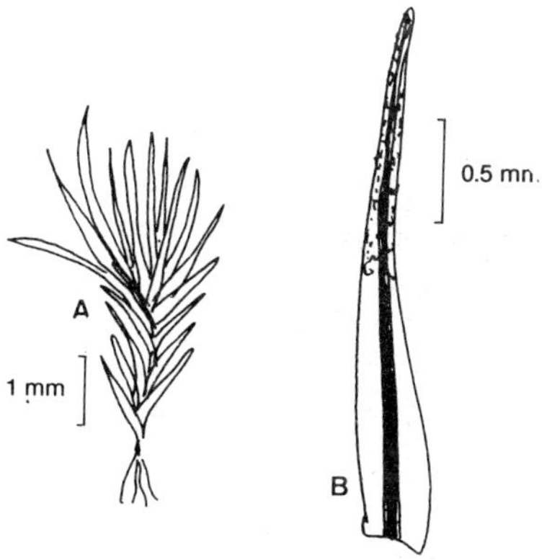

Fig. 54. VEILED-MOUTH POTT MOSS (*Hymenostomum edentulum*) A — whole plant; B — leaf.

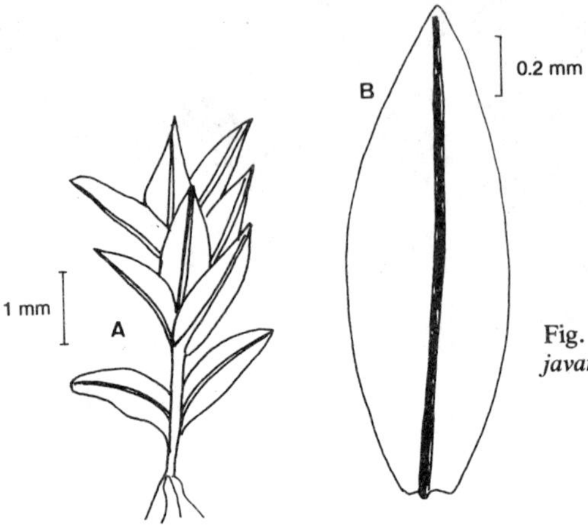

Fig. 55. JAVANESE POTT MOSS (*Hyophila javanica*) A — whole plant; B — leaf.

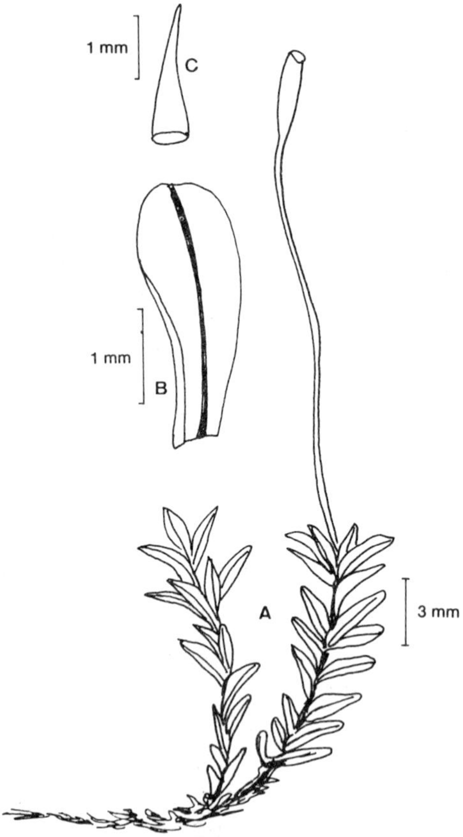

Fig. 56. INROLLED POTT MOSS (*Hyophila involuta*) A — whole plant; B — leaf; C — operculum.

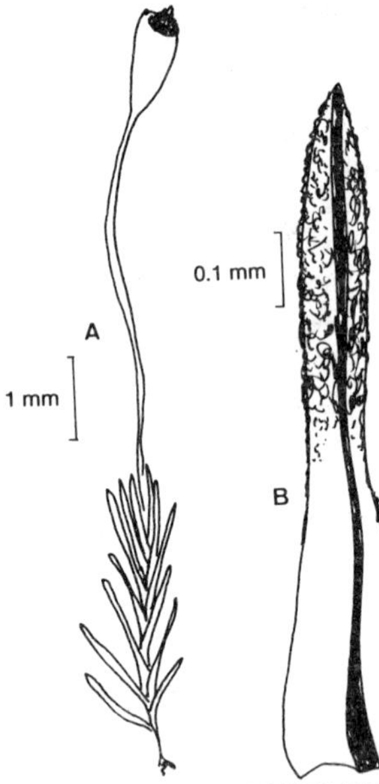

Fig. 57. EASTERN POTT MOSS (*Semibarbula orientalis*) A — whole plant; B — leaf.

Localities: Sandy soil and alluvial strip by sea, Kedah; Tapah, Perak; Kuala Pilah, Negri Sembilan; Waterfall Gardens, Penang; Jengka and Kuala Tahan, Pahang; Botanic Gardens, Nee Soon, Bukit Timah, and MacRitchie Reservoir, Singapore. Like *Fissidens crassinervis* this species can also tolerate full sunlight. It is widely distributed in Southeast Asia and Sri Lanka.

POTT MOSS FAMILY

(Pottiaceae)

(after Pott, once Professor of Botany in Brunswick)

The Pott Moss Family is an ill-defined family, consisting of small upright mosses which bear rhizoids at the base. It is poorly represented in this area, being commoner in temperate regions. The leaves are variable in shape and have a prominent nerve. They are often somewhat papillose. The seta is straight, bearing a small to minute, erect or slightly inclined capsule. The calyptra is usually hood-shaped and split down one side. The peristome consists of a single row of 16 teeth which are straight or spirally twisted, and may be divided into 32 long arms. Some of our local species are confined to limestone rocks.

SNOW-BORDERED POTT MOSS

Chionoloma latifolium (Fig. 53)

The Snow-Bordered Pott Moss is an upright leafy moss which occurs in tufts on limestone. The stem is single or rarely branched, 2−3 cm high, and may be bent upwards from a short creeping portion bearing numerous red rhizoids. The leaves are very thickly arranged on the stem, brittle, and falling off below. When dry they are curled up, crisped, and falcate. When wet and spreading they form a more or less regular terminal rosette. They are lanceolate, usually bent along the nerve. The tip is pointed with an excurrent nerve. The lower half of the leaf has a prominent snow-white border of elongated, rectangular, hyaline cells. The upper lamina cells are square or somewhat elongated near the nerve, and strongly papillate. The white border cells are smooth, not papillate. Basal lamina cells are much larger, elongate, smooth, and without papillae. The capsule has not been seen.

Localities: Pulau Dayang Bunting, Langkawi.

VEILED-MOUTH POTT MOSS

Hymenostomum edentulum (Fig. 54)

The Veiled-Mouth Pott Moss is a tiny olive green plant, occurring in thick tufts 2—10 mm high. The stem is upright, often branched at the base. The leaves are longer and thicker towards the top of the stem, and when dry they are folded. They are lanceolate awl-shaped, with an acute tip. The upper leaf cells are rounded and strongly papillose, so that it is difficult to see cell walls. The nerve has a hyaline tip which is bent abruptly. The leaf margin is crenate and papillose. A border is present in the upper part of the leaf. Basal leaf cells are rectangular and smooth.

The seta is yellow, about 5 mm high. The vaginula is long ovoid. The capsule is ovoid. No peristome is present but the mouth is closed with a veil when young. The operculum is beaked and the calyptra hood-shaped with a single lateral split.

Localities: Pulau Langkawi near Kuah, on the roots of *Pyrrosia stigmosa* on limestone rocks near the sea. Widely distributed in Indonesia Sri Lanka, India, and the Philippines.

JAVANESE POTT MOSS

Hyophila javanica (Fig. 55)

The Javanese Pott Moss is a small plant about 5 mm high, forming thick tufts often matted with mud. The stem is single or rarely branched from the base and red-brown in colour. The leaves are ovoid to tongue-shaped, and borne in a rosette at the top of the stem. The margin is smooth, the apex acute, and the nerve prominent of a pale brown to golden colour and not quite reaching the tip. The upper laminal cells are hexagonal and transparent. They gradually lengthen below to a rectangular shape and become lacking in chloroplasts.

The perichaetial leaves are very small, the inner ones without a nerve. The seta is 7 mm high, yellow or red in colour, and bears an upright cylindrical capsule. Stomates may be found on the neck. The operculum is about half the length of the capsule, and is long beaked. The annulus is not prominent. There are no peristome teeth.

Localities: On earth and walls at Fraser's Hill and in Perlis. Found also in Indonesia.

INROLLED POTT MOSS

Hyphila involuta (Fig. 56)

The Inrolled Pott Moss is a small upright moss about 1.5 cm high. The primary stem is erect and usually once branched. The upper and perichaetial leaves form a teminal rosette. Upper leaves are lanceolate-linear, with the upper margins inrolled. Lower leaves are more or less ovate, with a rounded tip and entire margins. Upper cells are opaque and covered by small warts, while lower cells are smooth and rectangular, the marginal ones being smaller than the inner ones. The nerve is prominent. It occupies about one-tenth of the width of the leaf base and extends to the tip. It is not excurrent.

The seta is long and very slender. The capsule is narrow cylindrical, inclined or rarely upright. The annulus is prominent, red in colour, and rolled over. The operculum is beaked. There are no peristome teeth, but inside the rim at the edge of the mouth there is a row of swollen white lamellae.

The base of the plant may be clothed in a mass of red paraphyllia.

Localities: Found growing directly on limestone in Langkawi. It is one of the few mosses growing directly in limestone in this area.

*EASTERN POTT MOSS

Semibarbula orientalis (Fig. 57)

The Eastern Pott Moss is a small insignificant moss growing in loose tufts on the ground. The stems are only 5 mm high, unbranched, or occasionally branched from the base. The leaves are rather distant from one another, borne at an angle to the stem, and clustered at the top. They are broadly lanceolate with a small excurrent tip, and often folded along the nerve. The margin is crenate in the upper half. Upper lamina cells are very small, rounded, and papillose, and completely opaque. The lower lamina cells are rectangular and smooth. The nerve is very prominent, papillose on the dorsal side, and biconvex.

This species is nearly always fertile. It has a prominent long cylindrical vaginula. The seta is 10 mm high, yellow at the top and red at the base. The capsule is ovoid with a wide mouth. The peristome is single. Each tooth is split into two quite separate filiform, papillose parts.

Localities: This is very common on bare ground in the sun, especially in Singapore and Penang. It is also found in Indonesia and Sri Lanka.

LOUISIADEN POTT MOSS

Barbula louisiadum (Fig. 58)

The Louisiaden Pott Moss is an upright moss growing in loose tufts about 1 cm high. It is pale green in colour, with a pale stem which is unbranched and clothed with numerous pale lanceolate leaves. The leaves have long pointed tips with a hyaline end cell. The nerve is bright yellow in colour, except in the oldest leaves where it turns red. The margin is crenate towards the tip. Upper lamina cells are opaque and papillose; while lower cells are rectangular, not papillose. The nerve is slightly nodulose on the dorsal side.

Localities: Penang and Perlis. Also found in Louisiaden and the Carolines.

***CURVED-BEARD POTT MOSS**

Hydrogonium arcuatum (Fig. 59)

The Curved-Beard Pott Moss is a small tufted plant about 3 mm tall. The leaves are long, triangular, and twisted when dry. They have a prominent excurrent nerve. The base of the plant may be trailing. The capsule is erect, on a long seta about 3 mm high. It is long cylindric, bearing remarkable spirally twisted peristome teeth, covered by a conical operculum. The teeth are supposed to resemble a small beard.

Localities: This has been recorded from Singapore.

SARAWAK POTT MOSS

Trichostomum sarawakensis (Fig. 60)

The Sarawak Pott Moss is an upright plant about 2 cm high. The stem is usually unbranched, and red at the base. The leaves are lanceolate-linear, with a sheathing base, and folded along the nerve. The apex forms a small tip. The leaves are opaque, held at an angle from the stem, and crisp when dry. They are pale-green with a reddish mid-rib, especially in the older leaves. The leaves are distant at the base. Basal leaf cells are elongated and hyaline. Upper leaf cells are small, rounded, and strongly papillose. There is no border.

Localities: On limestone crevices in Perak and Sarawak.

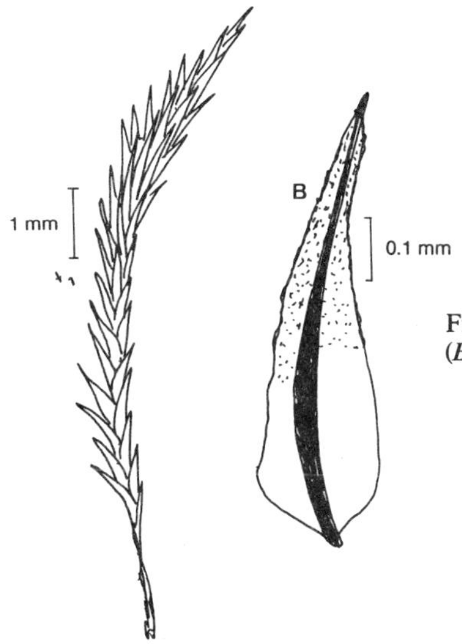

Fig. 58. LOUISIADEN POTT MOSS (*Barbula louisiadum*) A — whole plant; B — leaf.

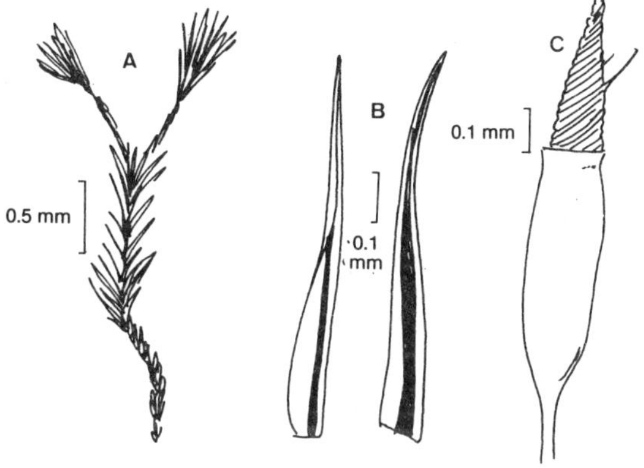

Fig. 59. CURVED-BEARD POTT MOSS (*Hydrogonium arcuata*) A — whole plant; B — leaf; C — capsule.

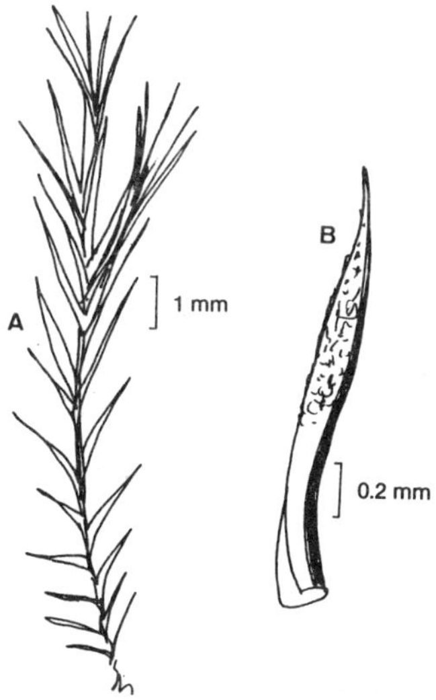

Fig. 60. SARAWAK POTT MOSS (*Trichostomum sarawakensis*) A — whole plant; B — leaf.

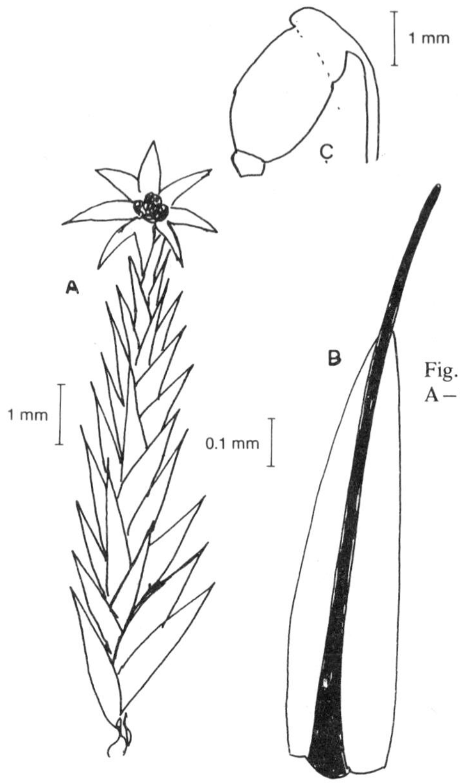

Fig. 61. CROWN MOSS (*Bryum coronatum*)
A – whole plant; B – leaf; C – capsule.

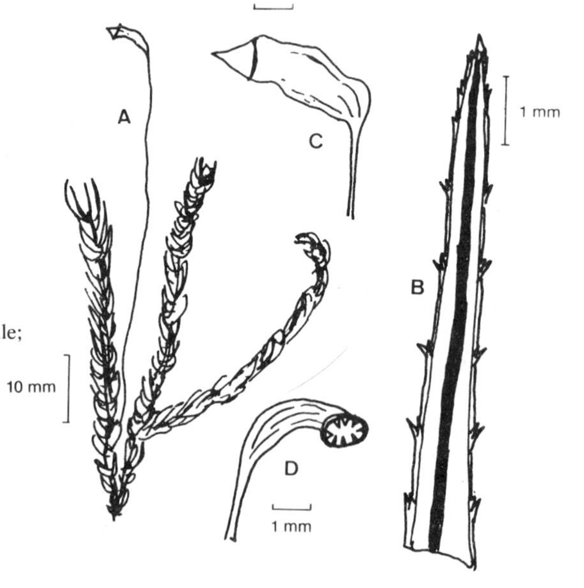

Fig. 62. SPINY BRUSH MOSS
(*Rhizogonium spiniforme*)
A — whole plant; B — leaf; C — capsule;
D — capsule without operculum.

BRYUM FAMILY

(Bryaceae)

Mosses in the Bryum Family form tufts with the stem sometimes branching below by innovation. The leaves increase in size from the bottom upwards to form a comal tuft above. The nerve is long and excurrent. The lamina cells are usually large, with thin oblique walls. The calyptra is narrow, hood-shaped, and soon falls off. The capsule is pendulous on an elongated seta or, rarely, inclined or erect. It is variable in shape. The operculum is convex. The peristome is double, with 16 outer lanceolate undivided teeth which are thickened with horizontal bars; plus an inner peristome of 16 processes alternating with the teeth, with or without cilia present. This family is poorly represented in the region.

*CROWN MOSS

Bryum coronatum (Fig. 61)

The Crown Moss forms tufts about 4 mm high, with erect ascending leaves. These are narrow ovate with a long excurrent nerve which is usually orange-red in colour and is toothed above. The lamina cells are long hexagonal. The margin is entire or slightly toothed.

The seta is red, 14 mm high, with an acorn-shaped capsule about 2 mm long. When ripe this capsule is dark brown to black in colour, and has a swollen neck. The peristome is double, with 16 outer teeth plus inner teeth with alternating cilia. The outer teeth are long, red, and lanceolate with strong horizontal thickenings. The inner teeth and cilia are whitish and delicate.

Localities: Common especially on walls, but becoming rare in Singapore.

◆ ◆ ◆

BRUSH MOSS FAMILY

(Rhizogoniaceae)

Mosses in the Brush Moss Family are upright mosses forming large, tall tufts looking like small bottle brushes. This is partly because of the dense spiral arrangement of leaves in the species described below, and partly because there is no coma at the top of the stem. The nerve is single, very prominent, and often toothed dorsally. There is also a prominent toothed border. The peristome is double, with the outer peristome shorter than the inner peristome.

*SPINY BRUSH MOSS

Rhizogonium spiniforme (Fig. 62)

The Spiny Brush Moss forms very loose, pale reddish-green tufts bearing dark, smooth cherry brown rhizoids at the base. The stem is upright, 1–5 cm high, sometimes quite stiff, often bend upwards from the base, usually simple, and seldom divided to form a rhizome or bulb-like structure. The leaves are quite narrow linear-lanceolate, with prominent excurrent toothed nerve. There is also a prominent narrow toothed border. All lamina cells are thick-walled, rounded, and green. Border cells are of the same shape but are deeply coloured and bear paired teeth.

Outer perichaetial leaves are smaller than the lower leaves hence there is no coma. The seta is 4 cm long, bearing an inclined cylindrical capsule 2 mm long. The capsule wall dries into slightly spiral folds. It has a wide mouth and a distinct neck.

Localities: Common throughout area, but rare in Singapore. Widely distributed in Asia, Africa, and tropical America.

*BROAD-LEAVED BRUSH MOSS

Rhizogonium latifolium

The Broad-Leaved Brush Moss species is generally smaller in size, with shorter leaves which are elliptical lanceolate and not, or scarcely, narrowed at the base.

Localities: Gunong Tahan, Pahang; Gunong Belumut, Johore: Penang Hill; Singapore.

◆ ◆ ◆

WEBER'S MOSS FAMILY

(Weberaceae)

The Weber's Moss family contains mosses which usually form small isolated tufts of narrow stems which are rarely branched above. The leaves may be broad or narrow below, but are exceedingly narrow in the coma. The nerve is prominent. The lamina cells are long and narrow. The capsule is often pear-shaped or cylindrical. The calyptra is small and soon falls off. The peristome is double, like that of the Bryum family, but without any cilia in the inner peristome, or rudimentary.

*WEBER'S SHORT-TIPPED MOSS

Diphyscium mucronifolium

The Weber's Short-Tipped Moss is a dark-coloured lowly plant with a few large spreading leaves at the base. The stem is scarcely 0.5 mm high. The lower leaves are about 3—5 mm long, tongue-shaped, with a short turned-down tip when dry. The nerve is prominent and stops just before the tip. The margin is entire. No border is present but the lamina is rather wavy, and two-cells thick. The perichaetial leaves are about 0.3 mm long, triangular shaped, and with a long excurrent tip about 0.5 mm long. The apex of these leaves may be two cells thick as is the margin. The rib is less distinct in the perichaetial leaves. The capsule has not been seen.

Localities: Gunong Keledang, Perak; Bukit Timah, Singapore. Found also in Indonesia.

——— ◆ ◆ ◆ ———

HAIR MOSSES

(Polytrichaceae)

Hair mosses are usually of fairly large to large size, and grow on earth or rocks in the mountains. The stem is usually simple and upright. The leaves are narrow and spirally arranged. The nerve is wide and usually produces upright columns of green cells along its upper surface (*lamellae*). These increase the photosynthetic area of the leaf.

The capsule is long cylindrical on a long seta. Its method of dehiscence is quite different from all other mosses described here. The teeth are small. numerous (32, 64, or 128) rigid structures which are firmly attached to a circular membrane (*epiphragm*) which closes the mouth. The teeth are not transversely barred and do not move in response to humidity. An operculum is present over the epiphragm, and the whole capsule is covered with an often hairy calyptra. When the calyptra and operculum fall off, the spores are dispersed by a "pepper pot" mechanism through the holes between the rigid teeth by shaking of the seta.

LARGE-LEAVED HAIR MOSS

Pogonatum macrophyllum (Fig. 63)

The Large-Leaved Hair Moss is a very common large moss found in the mountains above 1,000 m. It varies from about 4 cm to 35 cm in height. The stem is upright and wiry. It is seldom branched. The leaves are narrow, linear, with a slightly expanded basal part forming a sheath about one-fifth the length of the leaf. The nerve is prominent in the upper part but reduced below in the sheath. Unlike most other species of *Pogonatum,* it does not bear lamellae. The margin of the sheath portion is entire but it is toothed above. The capsule is borne on a long seta about 5 cm tall. The seta is wiry and often bent or curved. The cap-like calyptra is pale brown in colour and has longitudinal folds. The capsule is cylindrical with rigid teeth and an epiphragm.

Localities: Common in mountains of Malacca, Selangor, Pahang, Perak, and Penang.

JUNGHUHN'S HAIR MOSS

Pogonatum junghuhnianum

The Junghuhn's Hair Moss is a small hair moss about 1—2 cm high. It has firm stems often branched at the base. The toothed leaves stand out from the stem and are spirally arranged. Each leaf has a distinct sheath portion occupying one-quarter to one-half of the length of the leaf (or even more in perichaetial leaves). The nerve is prominent but contracted in the sheath portion. It bears long rows of lamellae.

The capsule is borne terminally on a long seta, 2—3 cm tall. The calyptra completely covers the young capsule and is fringed at the base. The capsule is urn-shaped and somewhat asymmetrical. It bears a short-tipped operculum. The peristome teeth are thick rigid rounded structures, attached to the epiphragm above.

Localities: Fraser's Hill.

RUINED-TOOTH HAIR MOSS

Racelopus pilifer (Fig. 64)

Male plants of the Ruined-Tooth Hair Moss are 2—3 mm high, with a basal swelling bearing long orange-red branched rhizoids. Main stem is simple or once branched and bears many erect-ascending leaves, the lowermost being reduced in size. The leaves surrounding the male head are triangular in shape and have a very long excurrent tip. Female plants are 2—5 mm high, somewhat larger than the male. The long, rough, red seta is 2 cm high. It is at first upright, covered with a shaggy calyptra. Later the calyptra is shed and the capsule assumes a horizontal position.

Although mosses have a strong dislike of salt water and few are found near the sea, this species is exceedingly common on the sandy beaches of the east coast of West Malaysia. It appears to act as a sand-binder, consolidating the unstable sand, previous to the grass invasion

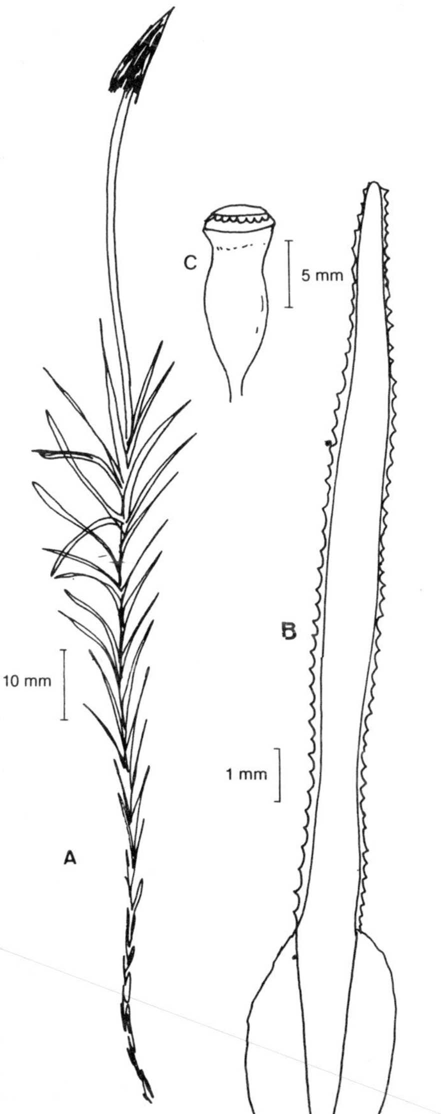

Fig. 63. LARGE-LEAVED HAIR MOSS (*Pogonatum macrophyllum*) A — whole plant; B — leaf; C — capsule.

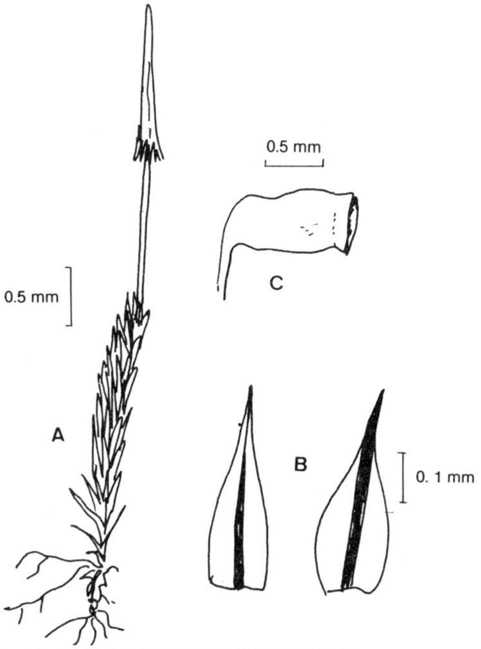

Fig. 64. RUINED-TOOTH HAIR MOSS (*Racelopus pilifer*) A — whole plant; B — leaves; C — capsule.

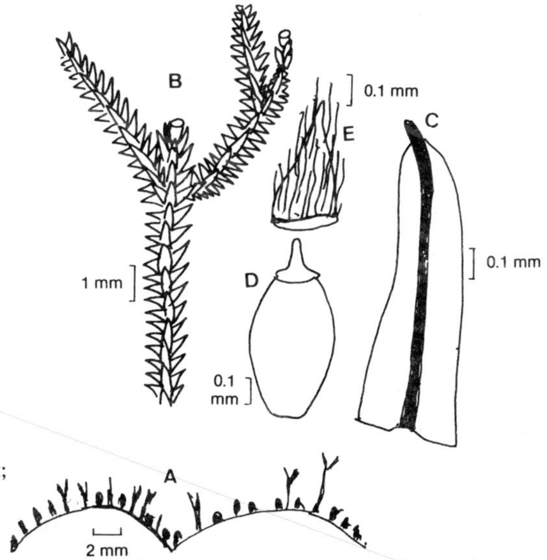

Fig. 65. POINTED HIDDEN-FRUIT MOSS (*Desmotheca apiculata*) A — whole plant; B — fertile branch; C — leaf; D — capsule; E — calyptra.

stage. Different sexes grow in separate patches several inches in diameter, making fertilization a matter of chance. It is also found on exposed rocks and earth in the mountains.

Localities: Gunong Tampin, Negri Sembilan; Ginting Peras and other hills, Selangor; Fraser's Hill, Pahang; Maxwell's Hill and other hills, Perak; sandy beaches at Kuala Sedili, Dungun, Kuantan, Kuala Trengganu, Kuala Kerai, and Kota Bahru. Found also in Thailand and Indonesia.

◆ ◆ ◆

STRAIGHT HAIR MOSSES

(Orthotrichaceae)

The Straight Hair Mosses tend to creep over the surface of bark or rock, and have turned-up branch tips to give dense cushions. The leaves are oblong-lanceolate to linear-lanceolate in shape. They are completely curled up when dry. The lamina consists of rounded hexagonal cells usually covered with numerous small papillae.

The seta is formed at the ends of the branches and is short (or very short). When the seta is very short so that the capsule is buried among the perichaetial leaves, it is said to be *immersed.* The capsule is always erect and quite symmetrical, with a smooth or striate wall. It is completely covered with the calyptra which may be hood- or bell-shaped, and usually bears straight hairs. The peristome may be single or double and is rarely absent. If present, the outer peristome teeth may be grouped in pairs.

*POINTED HIDDEN-FRUIT MOSS

Desmotheca apiculata (Fig. 65)

The Pointed Hidden-Fruit Moss is a creeping plant with rigid upright branches, 5—15 mm tall (sterile); 20—30 mm tall (fertile). The tufts are orange to rust coloured. Branch leaves are closely crowded and strongly papillose with small rounded papillae near the apex, but larger smooth cells at the base of the leaf. The margin is slightly toothed, especially at the base. The main stem leaves are broadly ovate and very crowded.

Fertile branches are very different from sterile ones. The leaves are borne strictly in four rows and are ovate with an extended tip. The capsule is hidden at the top of these branches among the perichaetial leaves. The seta is very short. The capsule has a swollen base with a prominent annulus above. There are no peristome teeth. The operculum has an abrupt tip. The calyptra is cap-shaped, covered with long hairs.

Localities: Bukit Timah, Singapore. Also found in Indonesia, the Philippines, and Burma.

*CURVED-LEAF STRAIGHT HAIR MOSS

Macromitrium incurvifolium

The Curved-Leaf Straight Hair Moss is a slender plant with a golden colour found on the bark of trees. The branches are elongated and clothed with lanceolate to tongue-shaped leaves which have entire margins. When this plant is dry the leaves are incurved or inrolled like a snail's shell. Each leaf is about 0.2 mm long. The cells of the lamina are extremely small, rounded, or square-shaped, faintly papillose above and smooth below.

The seta is short, bearing an ovoid capsule with a wide mouth. The peristome teeth are ill-defined and lanceolate in shape.

Localities: Singapore.

◆ ◆ ◆

NECKER'S MOSS FAMILY

(Neckeraceae)

The Necker's Moss Family consists of mosses found on tree bark which have twining main stems which bear secondary pinnate or bipinnate stems. These secondary stems may also be creeping; or may be erect, borne at an angle to the tree branch or trunk; or may hang down away from the trunk. The secondary stems bear leaves arranged in a single plane (*complanate*). The leaves are often ovate-lanceolate or tongue-shaped. The nerve may be short or completely absent. The lamina cells, which are pale green in colour, are small rhomboid or linear in shape.

The vaginula is very prominent at the base of the seta which may be very short and almost absent in *immersed* species, or much longer in *exerted* species. The capsule is always erect and symmetrical with a single or double peristome.

ELEGANT INNER-HAIR MOSS

Endotrichella elegans (Fig. 66)

The Elegant Inner-Hair Moss is a glossy plant with a creeping primary stem, bearing only a few leaves but many rhizoids. This bears irregular upright flattened branches, about 5—10 cm long, bright yellow green in colour, and sometimes branched again. The leaves are rather loose, more or less regularly overlapping at the base and twisted in the upper half when dry. Each leaf has a number of longitudinal folds, being ovate or ellipsoidal below, and drawn out to a long tip above.

The capsule is hidden among the perichaetial leaves. The seta is short, not more than 2 mm long. The capsule is long cylindrical and has a distinct ring round the mouth. There are two rows of peristome teeth. The outer teeth are bent outwards in the dry capsule, resembling a flower. Each tooth is transversely barred, and is divided along the median longitudinal line. The inner teeth are colourless, consisting of long segments on a basal membrane. The operculum covering the teeth has a long abrupt point. It is covered with a conical calyptra which splits down one side when young.

Localities: This moss is found in the mountains in Pahang and Perak. It is widely distributed in Indonesia, Taiwan, the Philippines, and the Caroline Islands.

NECKER'S JOINED-TOOTH MOSS

Symphysodon neckeroides (Fig. 67)

The primary stem of the Necker's Joined-Tooth Moss is branched and creeping, often devoid of leaves. The secondary stems are stiff, much branched, and about 3 cm high. They are devoid of leaves at the base, but bear regularly arranged, concave, single-nerved leaves in six apparent rows above. These secondary stems are flattened in a single plane. Each leaf is triangular to tongue-shaped, about 3 mm long, and ends in a distinct point.

The capsule is hidden among the perichaetial leaves, and has a very short seta. The operculum arises deeply in the mouth of the capsule but is elevated later. The outer peristome teeth are found in eight pairs and are deeply inserted.

Localities: Gunong Raya, Langkawi. Widely distributed in Indonesia and the Philippines.

FEATHERY STRAP-SHAPED MOSS

Himantocladium plumula (Fig. 68)

The primary stem of the Feathery Strap-shaped Moss is rhizome-like, bearing few leaves but many rhizoids. The secondary stems are more or less regularly pinnately branched, 4–6 cm long. The leaves are dull yellowish green in colour, and are closely overlapping. They are tongue-shaped with a wider base and a wavy margin, especially when dry. The nerve is quite prominent, ending a short distance from the apex.

The seta is short, about 1.5 mm long. The capsule is urn-shaped, and partly obscured by the perichaetial leaves. The operculum is deeply

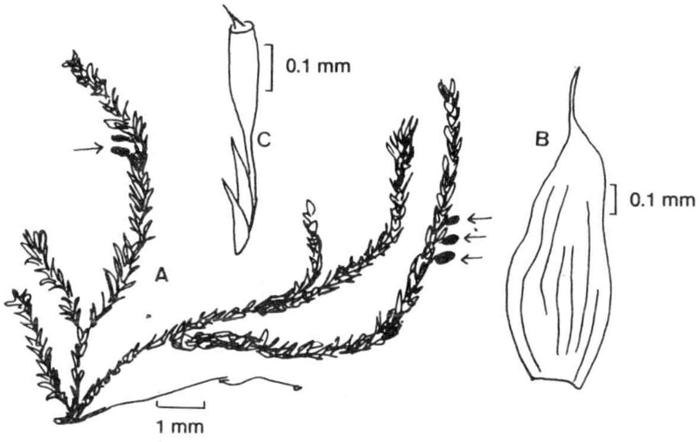

Fig. 66. ELEGANT INNER-HAIR MOSS
(*Endotrichella elegans*)
A — whole plant, with fruit arrowed;
B — leaf; C — capsule.

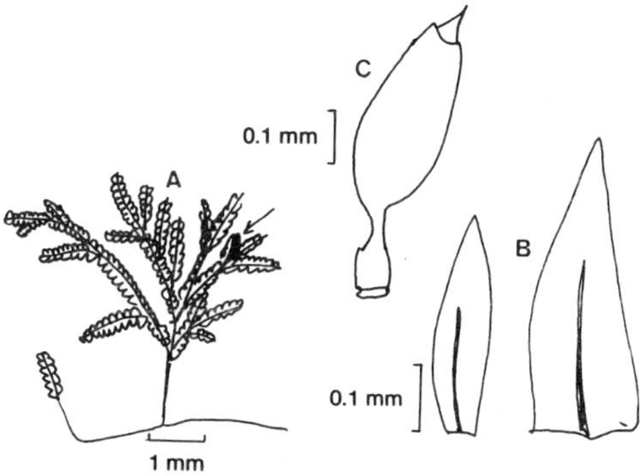

Fig. 67. NECKER'S JOINED-TOOTH MOSS
(*Symphysodon neckeroides*) A — whole plant;
B — leaves; C — capsule.

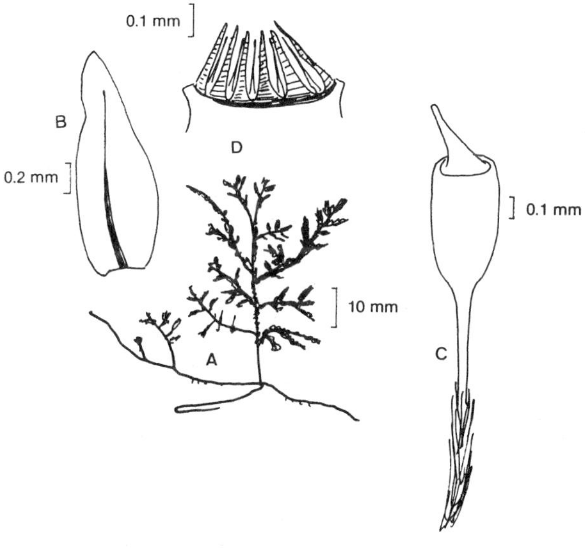

Fig. 68. FEATHERY STRAP-SHAPED MOSS
(*Himantocladium plumula*) A — whole plant; B — leaf;
C — capsule; D — peristome teeth.

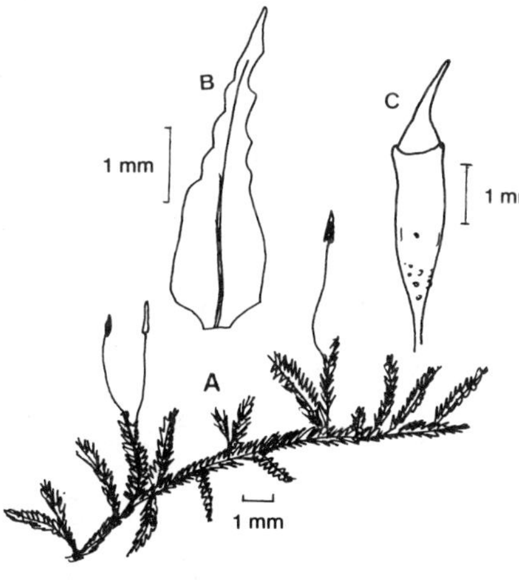

Fig. 69. ELONGATE AERIAL MOSS
(*Aerobryopsis longissima*) A — whole plant;
B — leaf; C — capsule.

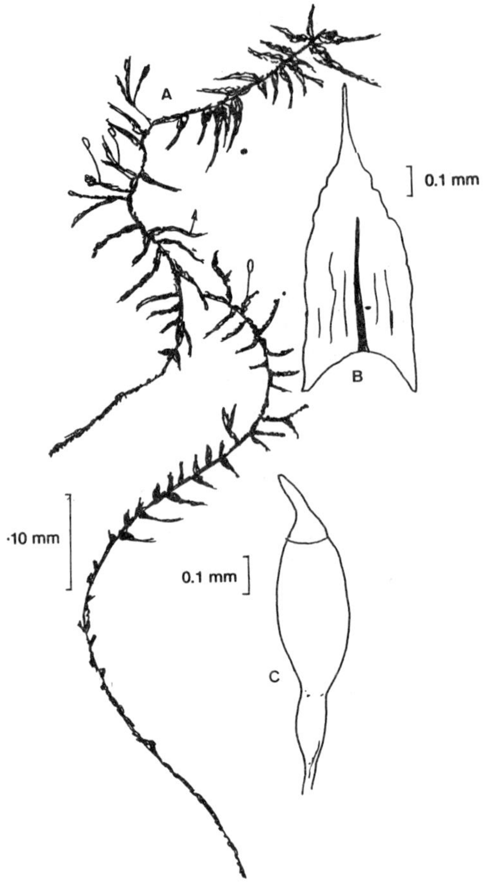

Fig. 70. MIQUEL'S MOSS (*Meteorium miquelianum*) A — whole plant; B — leaf; C — capsule.

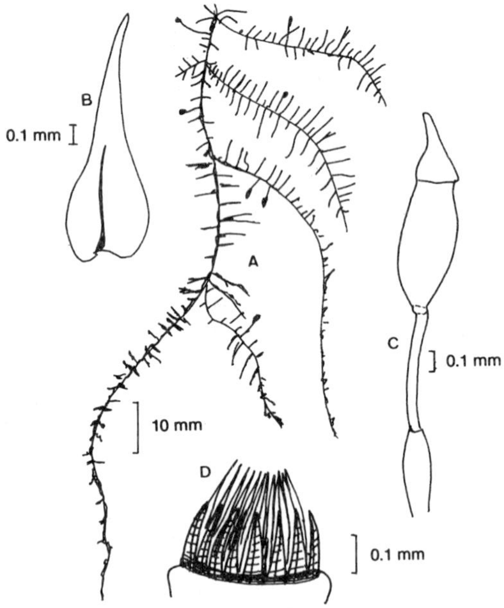

Fig. 71. PROFUSELY FERTILE MOSS (*Floribundaria floribunda*) A — whole plant; B—leaf; C—capsule, D—peristome.

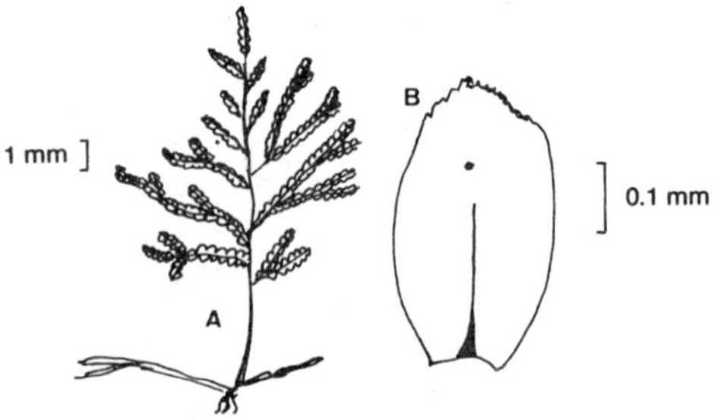

Fig. 72. SMALL MUCRONATE FEATHER MOSS (*Pinnatella mucronata*) A — whole plant; B — leaf.

inserted in the mouth with a cap-like base and an asymmetrical pointed tip. There is a double peristome. The vaginula is covered by hairs.

Localities: Usually epiphytic on trees, with primary stem winding round the trunk and the secondary stems set out at an angle. Recorded from Malacca, Selangor, Trengannu, and Perak. Also found in Indonesia, the Philippines, and New Caledonia.

ELONGATE AERIAL MOSS

Aerobryopsis longissima (Fig. 69)

The Elongate Aerial Moss is a long, shiny plant of a pale yellowish green colour, often with a brownish tinge. The secondary stems hang down from the primary stems, and are long with irregular branches. All the branches are flattened with wide, spreading leaves. Each leaf is roughly triangular but is abruptly narrowed at the base. The tip is long extended and is transversely wavy.

The seta is long and slender, with a narrow capsule bearing warts on its neck. The operculum is wider than the capsule at the base, but no prominent annulus is present. The operculum shape is wide conical.

Localities: This rather variable species is found throughout the area. It is most common on tree trunks in elevated forests. It is also found in India, Sri Lanka, China, and the Pacific Islands.

MIQUEL'S MOSS

Meteorium miquelianum (Fig. 70)

The Miquel's Moss is a magnificant branching moss with a small primary stem clinging to the tree trunk. The secondary stems are more or less pinnate and densely covered with leaves. Each branch is swollen at the base and resembles a small cone. The leaves are very concave, and have distinct longitudinal folds when dry. They are heart-shaped at the base with an irregularly wavy or infolded margin. The leaf tip is drawn out to a sharp point.

The seta is 3—4 mm long, curved, and usually smooth. The capsule may have a distinct neck, and is roughly ovoid in shape, apart from the neck. The lid is small and wide pointed, with an asymmetrical tip.

Localities: This species is recorded from Gunong Berumban, Pahang; and Batu Caves, Selangor. There is also a variety (var. *turgifolium,* turgid-leaved) found in the Cameron Highlands. However, I believe it is common throughout the country. It is also found in North India, Sri Lanka, Indonesia, the Philippines, China, Japan, and Fiji.

PROFUSELY FERTILE MOSS

Floribundaria floribunda (Fig. 71)

The Profusely Fertile Moss is a slender feathery plant of a dull yellowish or pale green colour. The main stem is about 20 cm long and bears laxly erect, feathery, bipinnate branches. Both the main and branch stems bear numerous small leaves. Stem leaves are longer with a long point, while branch leaves are broader. The plant is irregularly flattened.

The seta is 3 mm long, bearing a tiny 2 mm long capsule. The lid is wide conical with a broad tip. There is a well-marked annulus and a distinct double peristome. As the name implies, numerous capsules are formed.

Localities: Found in Batu Caves, Selangor; Perak; and Penang. It may be found on twigs and leaves in damp forests. Probably common throughout the country.

*SMALL MUCRONATE FEATHER MOSS

Pinnatella mucronata (Fig. 72)

The Small Mucronate Feather Moss is a lustreless, yellowish plant with a long creeping stem which is almost devoid of leaves but bears numerous rhizoids. The secondary branches are flattened and bipinnate. Leaves are found on the secondary branches. These leaves are ellipsoid in shape with an asymmetrical nerve reaching three-quarter way up the leaf. Each leaf is concave, somewhat toothed near the apex, and with a short pointed tip. Lamina cells are rounded above, but more elongated below. Each cell has one inconspicuous papilla.

Localities: Found in Singapore; Kota Tinggi, Johore; Gunong Tahan, Pahang; and Tapah, Perak.

◆ ◆ ◆

HOOKER'S MOSS FAMILY

(Hookeriaceae)

(After W.J. Hooker, renowned botanist, and author of many floras)

The Hooker's Moss Family is a family of creeping mosses found on tree bark. Both primary and secondary stems are creeping and bear spirally arranged leaves arranged in a single plane. The leaves are usually wide at the base, and ovate or lingulate in shape. Each leaf usually bears two short nerves, but they are very inconspicuous. The margin is often distinctly bordered with elongated lamina cells. The cells of the main part of the lamina are large and have thin walls.

The capsule is borne on a long seta and is usually inclined. The peristome is double; the inner peristome has a basal membrane. The calyptra is shortly lobed at the base.

YELLOW BENT-TOOTH MOSS

Campylodontium flavescens (Fig. 73)

The Yellow Bent-Tooth Moss forms extensive mats of relatively large, copiously branched, golden green, glossy plants. The stems are elongated and irregularly pinnate. Side branches are densely covered with leaves. These branches are about 1 cm long. Leaves are long ovate, with a long tip and with deep longitudinal folds. They are about 2 × 2.5 mm. No nerve is present. The entire lamina is made of very long, linear cells, except for basal angles. These angles form distinct auricles of almost square cells, different in colour from the rest of the leaf.

The seta is about 1.5 cm long, and yellowish or reddish in colour. The capsule is long cylindrical, about 2.5 mm long, and scarcely narrowed at the mouth. The outer peristome teeth are short, brown, and deeply inserted. They are easily bent or damaged. The inner peristome is rudimentary.

Localities: Found locally in Pahang and Penang. Widely distributed on tree trunks in India, Burma, Malaysia, Sri Lanka, Indonesia, and Austrialia.

POINTED TWO-ROW MOSS

Distichophyllum cuspidatum (Fig. 74)

The Pointed Two-Row Moss is a medium sized plant with flattened sparsely-branched stem about 1 cm high. The leaves are very contorted when dry, and are oblong with a rounded tip drawn out to a long point. Each leaf is about 3 mm long. From above two complete rows of leaves and two half-rows can be seen. There is a slight difference in size between the central and lateral leaves. Branched rhizoids are common at the base.

The seta is borne laterally, and is slender and curved, being about 5 mm long. The capsule is small and sometimes has a slight neck. The operculum is skull-cap shaped with a sudden narrow tip.

Localities: Penang Hill; Fraser's Hill. Distributed in Sri Lanka, Indonesia, and Pacific Islands.

SPOON-SHAPED TWO-ROW MOSS

Distichophyllum spathulatum (Fig. 75)

The Spoon-shaped Two-Row Moss has a long extended leafless base, and its leaves are regularly arranged above, with two complete and two half-rows visible on each side. The stem may branch several times. The leaves are rounded spoon-shaped, with a prominent mid-rib extending nearly to the tip. Unlike the previous species, the tip is not long drawn out.

The seta is borne laterally and is 7 mm long. The capsule is small, round, and inclined or hanging. The operculum is skull-cap shaped with a long or short tip. The calyptra completely covers the capsule and consists of a conical warty structure with a long fringe.

Localities: Cameron Highlands; Fraser's Hill.

*SLENDER-STEM TWO-ROW MOSS

Distichophyllum gracilicaule

The Slender-Stem Two-Row Moss is a slender plant with a creeping or low-lying stem with beautifiul, pale green succulent leaves. These leaves are spoon-shaped, with the widest part near the apex, and have a strong wide border. Laminal cells are large, wide, hexagonal shaped above, with lower ones three times as long as wide. The border consists of three rows of narrow, elongated cells. The leaf has a tiny tip which seems to be a continuation of the border. There is a not very prominent nerve reaching just over three-quarter way from the base to the tip of the leaf.

Capsule not seen locally. Elsewhere it is reported as having a smooth seta and calyptra.

Localities: Found on Bukit Timah, Singapore. Reported also from Indonesia.

*SCHMIDT'S TWO-ROW MOSS

Distichophyllum schmidtii (Fig. 76)

The Schmidt's Two-Row Moss is a robust plant with sparingly branched upright stems about 18 mm high. The leaves are large, roughly spoon-shaped, and with very wavy margins, especially when dry. There is a prominent border about two cells wide. Upper laminal cells are penta-

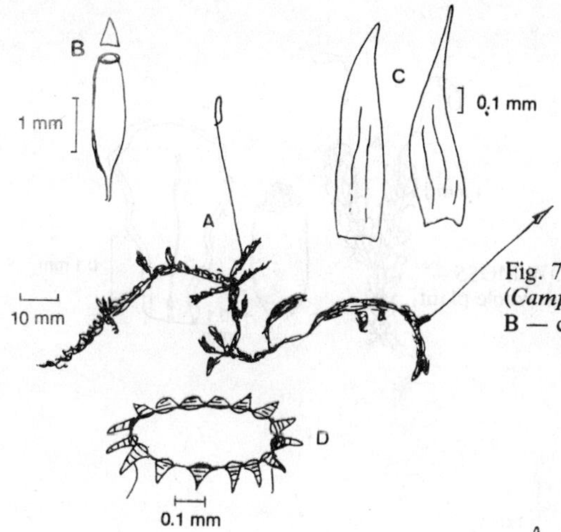

Fig. 73. YELLOW BENT-TOOTH MOSS (*Campylodontium flavescens*) A — whole plant; B — capsule; C — leaves; D — peristome teeth.

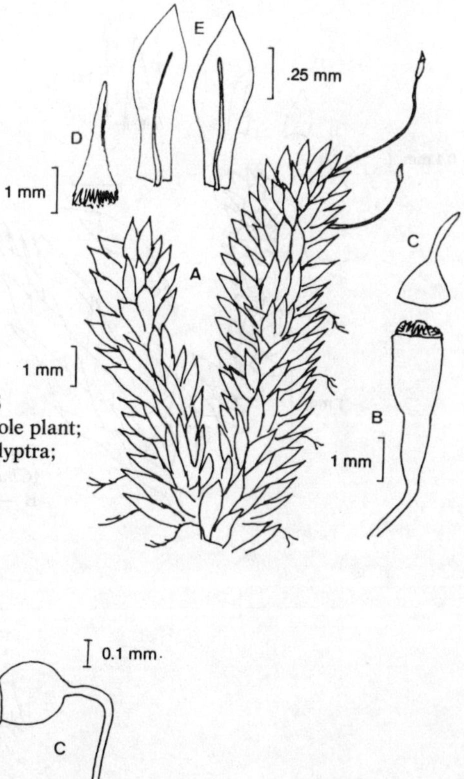

Fig. 74. POINTED TWO-ROW MOSS (*Distichophyllum cuspidatum*) A — whole plant; B — capsule; C — operculum; D — calyptra; E — leaves.

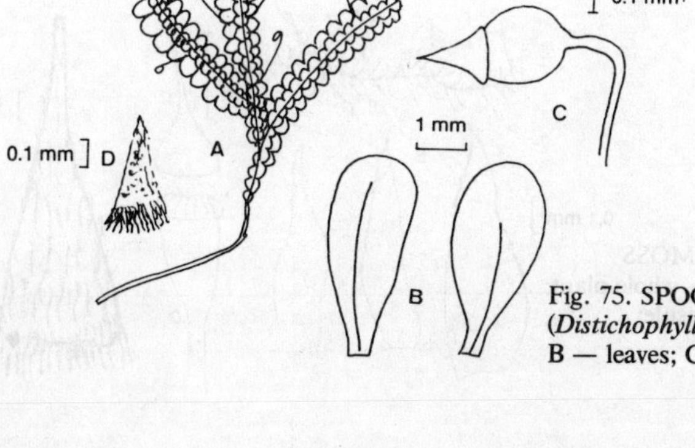

Fig. 75. SPOON-SHAPED TWO-ROW MOSS (*Distichophyllum spathulatum*) A — whole plant; B — leaves; C — capsule; D — calyptra.

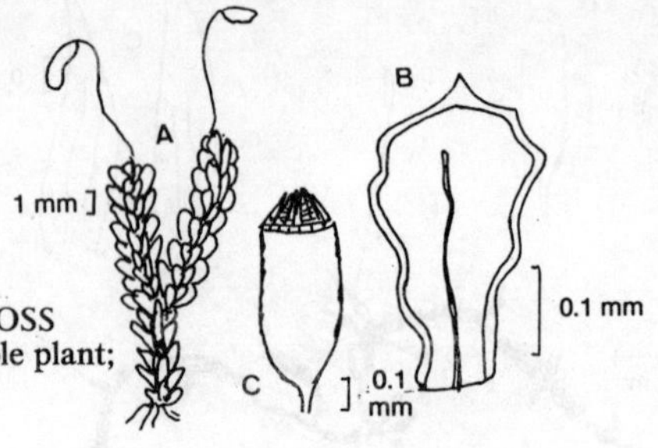

Fig. 76. SCHMIDT'S TWO-ROW MOSS (*Distichophyllum schmidtii*) A — whole plant; B — leaf; C — capsule.

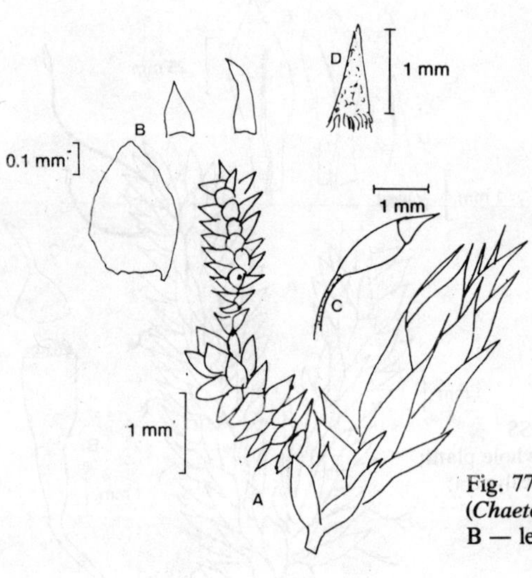

Fig. 77. SLENDER-FRUIT HAIRY-CAP MOSS (*Chaetomitrium leptopoma*) A — branch; B — leaf; C — capsule; D — calpytra.

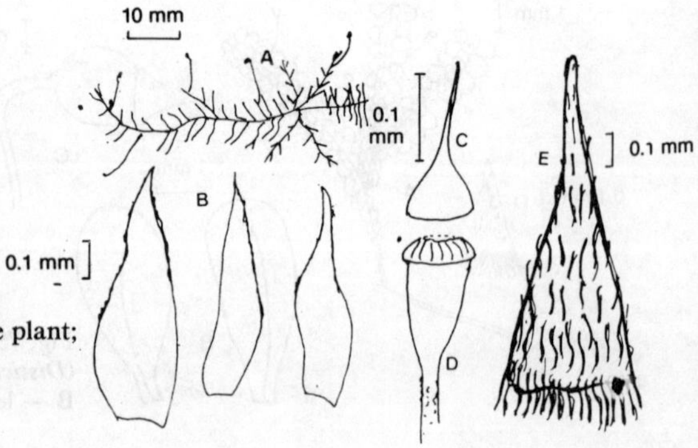

Fig. 78. ERECT-BEAK HAIRY-CAP MOSS (*Chaetomitrium orthorrhynchum*) A — whole plant; B — leaves; C — operculum; D — capsule; E — calyptra.

gonal or hexagonal in shape with lower cells lengthened and prismatic. The rib is prominent, wide, and wavy. It extends four-fifths to the tip. The apex has a small point which appears to be the extension of the border. Border cells are elongated.

The seta is about 8 mm long, with an inclined or horizontal capsule. The outer peristome teeth are very long and have very close bar-thickening. There is a prominent annulus. The outer teeth tend to adhere in pairs and stand erect when dry. Long filiform processes of the inner peristome can then be seen.

Localities: Bukit Timah, Singapore; Pulau Langkawi; Gunong Tahan. Also found in Thailand.

SLENDER-FRUIT HAIRY-CAP MOSS

Chaetomitrium leptopoma (Fig. 77)

The Slender-Fruit Hairy-Cap Moss is a small medium-sized creeping moss found on the bark of trees. The stem is long and creeping, bearing pointed triangular leaves. Secondary stems are irregularly pinnate, with concave leaves borne at a wide angle to the stem. The leaves have long drawn-out toothed tips.

The seta is surrounded by special clusters of perichaetial leaves on short side branches. It is long and covered with small papillae. The young capsule is completely covered with a warty calyptra, and with a fringe of long hairs at the base. The capsule is cylindrical, with a rounded operculum and a short tip.

Localities: Found in Selangor and Perak.

ERECT-BEAK HAIRY-CAP MOSS

Chaetomitrium orthorrhynchum (Fig. 78)

The Erect-Beak Hairy-Cap Moss is similar to the previous species but the branching is less regular. Also, the leaves are more regularly arranged. Each leaf has an ovate base drawn out to a wide or narrow toothed tip. The seta is very thick and papillose. The capsule is thimble-shaped, and may be upright or inclined. The operculum has a very long upright beak or point. The calyptra completely covers the capsule. It is hairy rather than warty.

Localities: Found in the Batu Caves and in Johore. Widely distributed in Indonesia.

ELONGATE HAIRY-CAP MOSS

Chaetomitrium elongatum (Fig. 79)

The Elongate Hairy-Cap Moss has a creeping stem with lateral branches that tend to arise on one side only. The leaves are very crowded and in many rows. Each leaf has a weird hooded apex. Main stem leaves have a semi-circular base and a wavy drawn-out tip, while secondary stem leaves are definitely tongue-shaped. Two short nerves may be seen.

The seta is borne on a special branch, with pointed perichaetial leaves. It is elongated and smooth. The capsule is pear-shaped, with a partly defined annulus and a double peristome. The operculum is rounded, with a long tip. The calyptra completely encloses the capsule when young. It is covered with long hairs and has its base split into segments.

Localities: Pahang; the Dindings. Distributed in Indonesia and the Moluccas.

*BEAUTIFUL-VEINED MOSS

Callicostella prabaktiana (Fig. 80)

The Beautiful-Veined Moss is a creeping moss with prostrate branches which are slightly flattened. It is readily recognized by the prominent double nerve with divergent branches ending near the apex. The leaves have long folds when dry and are somewhat variable in shape. The leaf margin is slightly crenulate. The seta is about 6 mm long, bearing a horizontally inclined capsule which has a swollen neck and a narrow constriction before the wide funnel-shaped mouth. The peristome is double. The calyptra is hairy and covers most of the capsule. It has a fringed base.

Localities: Singapore; Negri Sembilan; Penang; Perak; and Pahang. Probably common throughout the country. Also found in Indonesia.

◆ ◆ ◆

CONIFER MOSS FAMILY

(Thuidiaceae)

(from *Thuja*, botanical name for small feathery branched conifers, often used as Christmas trees in Singapore and Malaysia)

Mosses in the Conifer Moss Family are normally creeping, and are readily recognized by the regular bipinnate or tripinnate branching of the secondary stems. The primary stem is prostrate and irregularly divided. The much branched secondary stems often bear paraphyllia. The leaves on the secondary stems usually differ from those on the main stem;

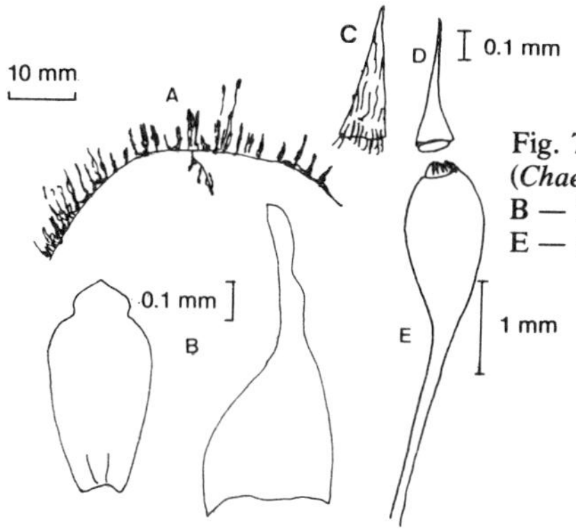

Fig. 79. ELONGATE HAIRY-CAP MOSS (*Chaetomitrium elongatum*) A — whole plant; B — leaves; C — calyptra; D — operculum; E — capsule.

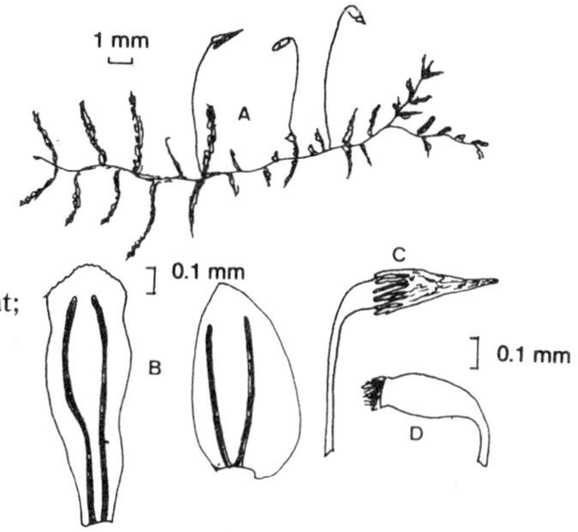

Fig. 80. BEAUTIFUL-VEINED MOSS (*Callicostella prabaktiana*) A — whole plant; B — leaves; C — capsule with calyptra; D — capsule.

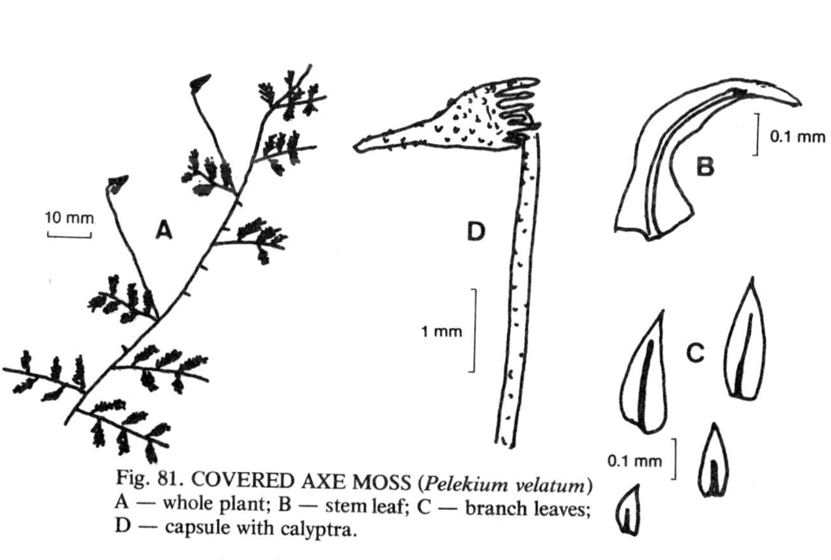

Fig. 81. COVERED AXE MOSS (*Pelekium velatum*) A — whole plant; B — stem leaf; C — branch leaves; D — capsule with calyptra.

branch leaves being smaller and well differentiated, ovate, concave, and short-tipped. All leaves are single nerved, with a lamina of numerous small highly papillose cells.

Capsules are usually horizontal or pendulous, but seldom erect. The seta is long and usually smooth. The peristome is double with both outer and inner peristome differentiated.

*COVERED AXE MOSS

Pelekium velatum (Fig. 81)

The Covered Axe Moss is a delicate lacy moss found throughout the country on logs, dead wood, tree bark, and rocks. The stems are wiry, bearing triangular-lanceolate leaves with a long and bent back tip. These stems bear numerous rhizoids and paraphyllia. The secondary stems are bipinnate, bearing broadly ovate, stumpy leaves. Both types of leaf have a prominent nerve, and all lamina cells are small rounded and very papillose. In branch leaves the nerve is very asymmetrical.

The capsule is usually horizontal and bears a prominent persistent calyptra which is bell-shaped, bearing a fringe of hairs at the base. These hairs stand out at angles from the calyptra. The capsule is urn-shaped, and has an operculum with a long slender beak. The seta is rather rough.

Localities: It is still found on the coral rockeries in the Botanic Gardens, Singapore. It has been recorded from Johore, Selangor, Negri Sembilan, Pahang, and Perak; and is probably widespread throughout Malaysia. Also found in Thailand, Pacific Islands, and the Philippines.

*TWO-ROW AXE MOSS

Pelekium bifarium

The Two-Row Axe Moss is a more robust but less common species than the Covered Axe Moss. The secondary branches are pinnate with the branch leaves spreading in two rows. These leaves are very small, being only 0.3 mm long at the ends of branches. They are oblong-shaped, with a rounded or blunt tip. The rib is prominent and often asymmetrical. The leaves become rolled up when dry. Lamina cells are small rounded and highly papillose. Stem leaves have long fine tips.

Localities: Singapore; Batu Caves, Selangor; Perak; Perlis; Gunong Tahan. Also in the Philippines and Indonesia.

HYPNON FAMILY

(Hypnaceae)

The Hypnon Family is a large family of creeping mosses. It is extremely variable in both branching and leaf shape. The nerve may be present or absent. The lamina cells are long and narrow, at least twice as long as wide, but usually much longer. They are usually smooth and without papillae. Alar cells may be differentiated but they are never inflated. The capsule is borne on an elongated seta. It is usually horizontal but, less commonly, may be erect. The double peristome is well developed. The inner peristome has a distinct basal membrane. Cilia may be present.

*NECKER'S OBLIQUE-FRUIT MOSS

Plagiothecium neckeroideum (Fig. 82)

The Necker's Oblique-Fruit Moss forms a thin mat on the ground, and is a glossy yellowish green in colour. The main stems are sparingly branched, and the whole plant looks somewhat untidy. The leaves are ovate or asymmetrical, about 2 × 1.2 mm. They are contracted at the base to form two auricles running down the stem on either isde. The nerve is double with two short, unequal forks.

The seta is about 2 cm long, and is strongly curved or bent at the top. The capsule is rounded, funnel-shaped, with a distinct neck. The outer peristome is rather short and stumpy.

Localities: Common and widely distributed. Also in Europe, Sikkim, Japan, and Java.

*BUITENZORG FOLDED-FRUIT MOSS

Ectropothecium buitenzorgii (Fig. 83)

(Buitenzorg was the old Dutch name for Bogor, Indonesia)

The Buitenzorg Folded-Fruit Moss is a common, robust, golden-green creeping moss which may be found on the ground or on trees. It forms very extensive mats. The stem is irregularly pinnately branched with the branches like the teeth of an irregular comb. The stem leaves are different from the branch leaves. They are triangular to lanceolate with a long falcate bent-over tip bearing minute teeth on the margin. Each leaf has two or three deep folds and several minor folds, but the nerve is completely absent. Branch leaves are ovate below, with or without a long falcate tip. All lamina cells are long linear, with a small group of wider square alar cells partially developed at the leaf corners. In stem leaves the alar cells may be larger and hyaline, but they are not inflated.

Perichaetial leaves are sharply toothed. The seta is about 2 cm long, bearing a pendulous ovoid capsule. The lid is convex with a short beak. The calyptra forms a folded hood split down one side.

Localities: Perhaps the commonest lowland moss found throughout Singapore and Malaysia. Also found in Indonesia and the Philippines.

*MORITZ'S FOLDED-FRUIT MOSS

Ectropothecium moritzii (Fig. 84)

The Moritz's Folded-Fruit Moss is a robust creeping moss, with a main stem more than 10 cm long. Secondary branches are regularly pinnate. Stem leaves are lanceolate with a wide toothed tip, usually with a single fold down the centre. There is no nerve but alar cells are prominent. Branch leaves are ovate, falcate, and also have an upper toothed margin. Alar cells consist of three or four enlarged cells bordered on each side by a few square cells. Normal lamina cells are rhomboid, about five times as long as wide.

Localities: Common and widely distributed.

*SINGAPORE FOLDED-FRUIT MOSS

Ectropothecium singapurense

The Singapore Folded-Fruit Moss is a creeping moss with an elongated primary stem bearing close, pinnately arranged, 5—7 mm long side branches. The leaves of the primary stem are compact, strongly falcate or curved, large, about 2 mm long by 1 mm wide at the base, and somewhat plicate. The shape of the leaves is roughly triangular. The margin is almost entire to weakly toothed. The cells of these leaves are small, narrow, short, and somewhat broadened towards the base. Branch leaves are smaller, narrower, more toothed, and bear a long, broad, acuminate tip. The upper cells of these leaves may be scabrous.

Localities: This has only been found in the Botanic Gardens, Singapore. It appears to be very rare.

*OBLIQUELY-INSERTED FOLDED-FRUIT MOSS

Ectropothecium incubans (Fig. 85)

The Obliquely-Inserted Folded-Fruit Moss is a slender, creeping, shiny moss with regular pinnate branching. Stem leaves are long

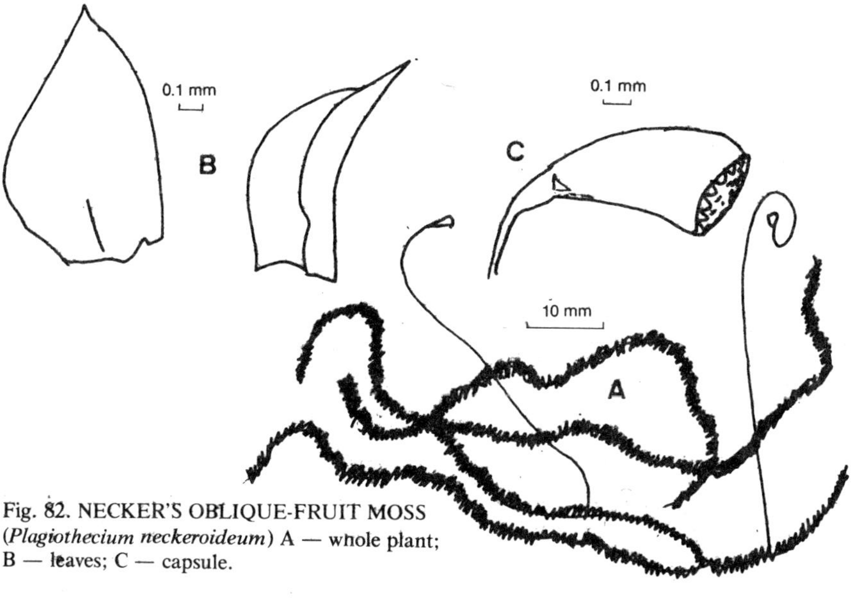

Fig. 82. NECKER'S OBLIQUE-FRUIT MOSS (*Plagiothecium neckeroideum*) A — whole plant; B — leaves; C — capsule.

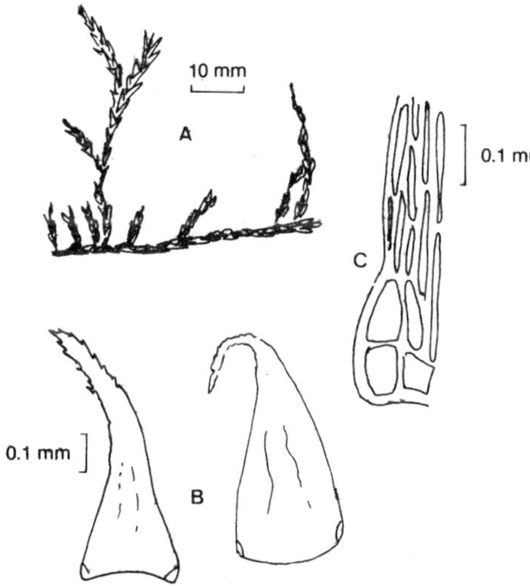

Fig. 83. BUITENZORG FOLDED-FRUIT MOSS (*Ectropothecium buitenzorgii*) A—whole plant; B — leaves; C — leaf corner to show alar cells.

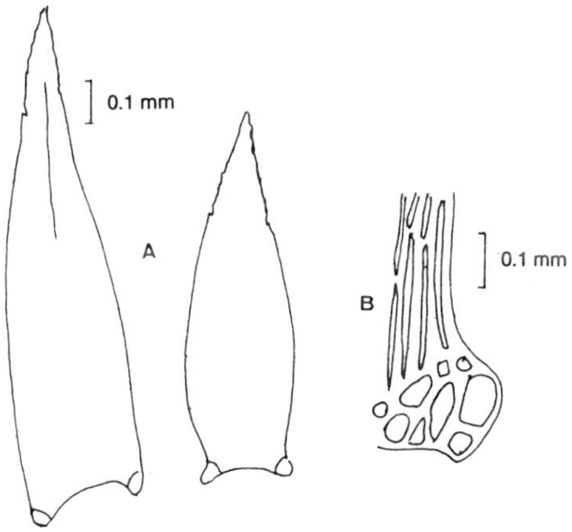

Fig. 84. MORITZ'S FOLDED-FRUIT MOSS (*Ectropothecium moritzii*) A — leaves; B — leaf corner to show alar cells.

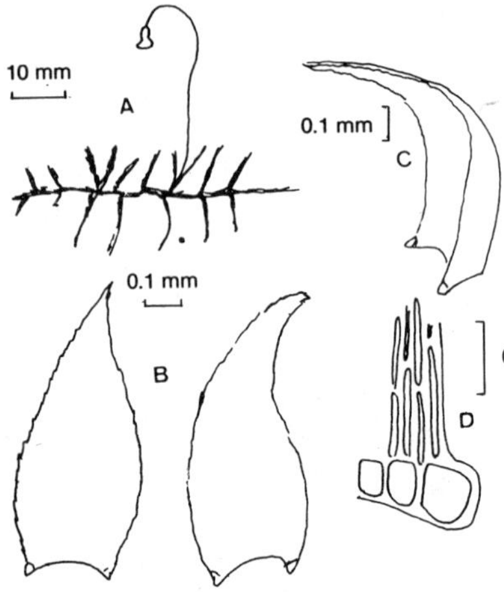

Fig. 85. OBLIQUELY-INSERTED FOLDED-FRUIT MOSS (*Ectropothecium incubans*) A — whole plant; B — leaves; C — side view of leaf to show concavity; D — leaf corner to show alar cells.

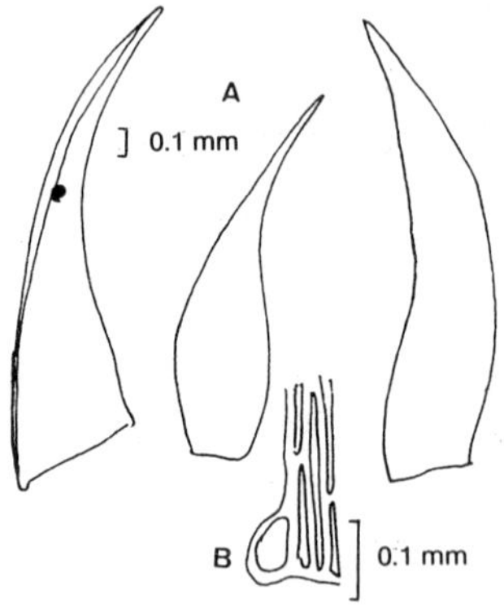

Fig. 86. MONUMENT FOLDED-FRUIT MOSS (*Ectropothecium monumentorum*) A — leaves; B — leaf corner with alar cell.

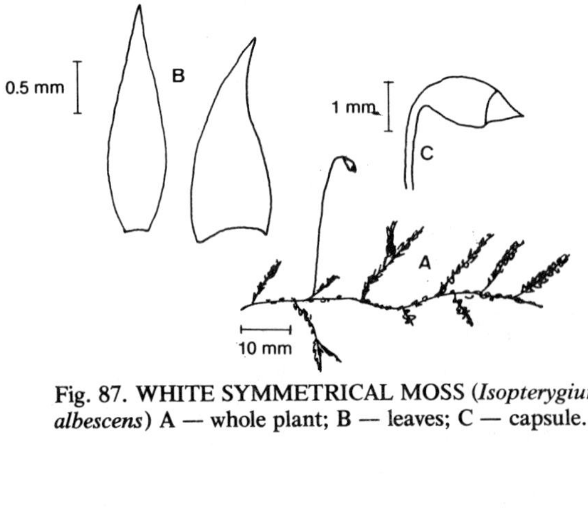

Fig. 87. WHITE SYMMETRICAL MOSS (*Isopterygium albescens*) A — whole plant; B — leaves; C — capsule.

falcate, concave, and folded. The elongated cells bear prominent papillae on their ends. The margin is toothed almost to the base. Basal angles bear three rows of large alar cells with very thin walls. Branch leaves are shorter, more ovate, and may lack teeth or be completely toothed. Alar cells are prominent.

Localities: Singapore; Negri Sembilian: Perak; Selangor. Also in Indonesia.

*MONUMENT FOLDED-FRUIT MOSS

Ectropothecium monumentorum (Fig. 86)

The Monument Folded-Fruit Moss is a small delicate species found on rocks and fallen logs. The stem is slender and more or less irregularly pinnate. It has a shining golden-yellow colour, and is quite flat. Stem leaves are slender lanceolate, with a few obscure teeth. Branch leaves are not much differentiated but are somewhat shorter. Cells are rhomboid, ten times as long as wide. There is one ill-defined alar cell at each leaf angle.

Localities: Bukit Timah and Botanic Gardens' Jungle, Singapore. Found also in the Philippines.

*MINUTE BRANCHED SYMMETRICAL MOSS

Isopterygium minutirameun

The Minute Branched Symmetrical Moss is a slender feathery moss of a brownish green colour. It forms thin mats on dead wood and wet rocks. The stems are creeping and irregularly pinnate. Stem leaves are somewhat asymmetrical. Branch leaves are ovate to lanceolate, spreading but not falcate, either entire or with small teeth near the tip. Lamina cells are narrow linear, and completely smooth. There is no nerve. No alar cells are differentiated. The branch leaves are very long and finely tipped.

The capsule is horizontal to pendulous. It is narrowly ovoid, and about 0.6 mm long. The operculum is sharply pointed.

Localities: Widespread throughout the country. Found also in Indonesia, Australia, the Pacific Islands, and the Philippines.

*WHITE SYMMETRICAL MOSS

Isopterygium albescens (Fig. 87)

The White Symmetrical Moss is pale whitish or silverish green in colour. It creeps along the ground to form thin mats. Stems are about 3 cm long with more or less regular pinnate branches. Branch leaves are broadly ovate, with long narrow tips.

Localities: Singapore, Selangor, Pahang, and Perak. Probably common throughout the country. Found also in the Himalayas, Sri Lanka, Sumatra. Indonesia, Japan, the Pacific Islands, and the Philippines.

*BANKA SYMMETRICAL MOSS

Isopterygium bancanum

The Banka Symmetrical Moss is a slender, flat, yellowish green, creeping moss. The leaves are long lanceolate, with a short tip usually serrated above. They are somewhat concave. There are no alar cells or nerve. The seta is 1.5 cm high, with a pendulous capsule bearing a conical operculum.

Localities: Batu Caves, Selangor; Penang Hill; Kroh, Perak; Singapore.

*MIQUEL'S BLADDER MOSS

Vesicularia miquelii

The Miquel's Bladder Moss, a robust, glossy, silvery green moss, forms lax mats on the bark of trees and exposed roots. The main stems are about 6 cm long and bear widespreading, pinnate branches which are about 6—8 mm long. The leaves are complanate, ovate, and long pointed, minutely toothed or entire near the tip, and about 1.3 mm long. The leaf cells are long and narrow. The seta is 2.5—3.0 cm long, slender, red in colour, and with an inclined or horizontal capsule with a conical lid and a short point.

Localities: Common and widely distributed in Singapore and Malaysia. Also found in Indonesia, the Philippines, and New Guinea.

*NET BLADDER MOSS

Vesicularia reticulata

The Net Bladder Moss is a dull, yellowish green, creeping moss found mainly on dead wood but sometimes occurring on soil. The main stem is about 5 cm long, with irregular flattened, pinnate branches. The leaves are spreading, ovate, long acuminate, and minutely toothed near the apex. They are about 1.5 × 0.6 mm. The leaf cells are long and rhombic in shape. The seta is 1.5—2.5 cm long, and bears a pendulous, ovoid capsule.

Localities: Bukit Panjang, Bukit Timah, Botanic Gardens, Singapore; Batu Caves; Maxwell's Hill; Penang; Kanga, Perlis; Tembeling; Kelantan. Also found in the Himalayas, Indonesia, the Philippines, Taiwan, and Oceania.

***DUBIOUS BLADDER MOSS**

Vesicularia dubyana (Fig. 88)

The Dubious Bladder Moss is an elongated, dull green, creeping moss with flattened irregular pinnate branches. The leaves are ovate, broadly acuminate, and short-pointed, and minutely toothed near the apex. They are about 1.3 mm long. Leaf cells are about eight times as long as wide and narrow rhombic in shape. The seta is about 1.5—2.00 cm long with ovoid, horizontal, or pendulous capsule.

Localities: Singapore; Batu Caves; Temengoh; Penang; Pulau Aor, Johore; Gunong Galas, Kelantan. Also found in Vietnam, Indonesia, Hong Kong, and Oceania.

***KURZ'S BLADDER MOSS**

Vesicularia kurzii

The Kurz's Bladder Moss forms loose, pale yellow-green shiny mats with a creeping main stem about 10 cm long, which bears long irregular branches. These branches, which are more or less pinnately arranged, are 4—10 cm long. The leaves are concave, oval, and acuminate. The stem leaves are blunter than the branch leaves. All leaves are distinctly toothed above. Leaf cells are thin-walled and of a long hexagonal shape. The seta is 1.00—1.5 cm high, reddish in colour, and bears a long ovoid capsule.

Localities: Bukit Timah, Singapore; Tanjong Malim; Tembeling. Also found in Indonesia.

***SPINY BEAKED MOSS**

Acanthorrhynchium papillatum (Fig. 89)

The Spiny Beaked Moss is a lustreless yellowish-green moss with elongate creeping stems forming extensive mats. The main stems have close pinnate side branches, which may be again pinnate. All branches

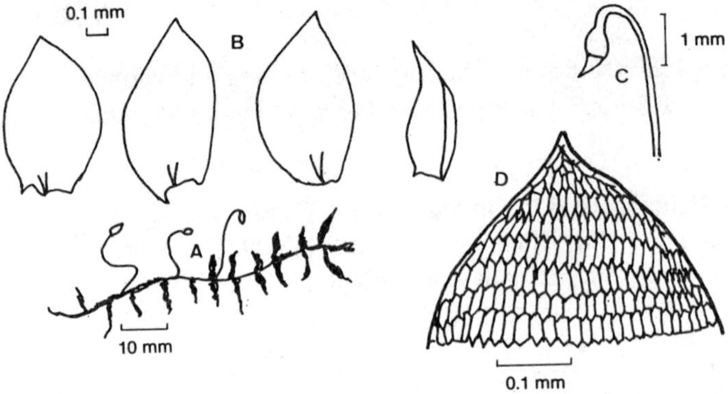

Fig. 88. DUBIOUS BLADDER MOSS (*Vesicularia dubyana*)
A — whole plant; B — leaves; C — capsule; D — leaf tip.

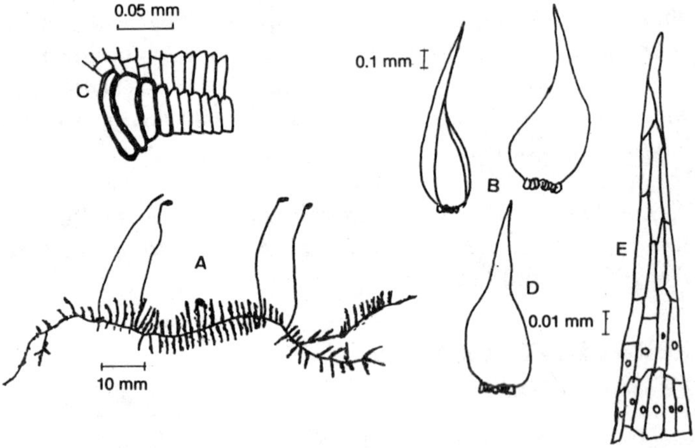

Fig. 89. SPINY BEAKED MOSS (*Acanthorrhynchium papillatum*)
A—whole plant; B—leaves; C—leaf corner to show alar cells;
D — leaf; E — leaf tip.

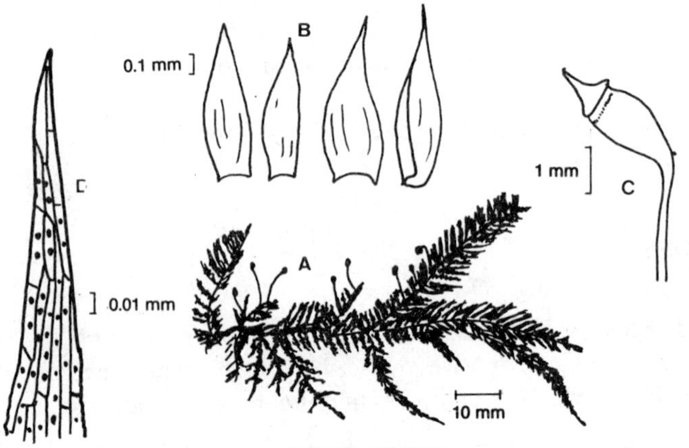

Fig. 90. KERI'S NIPPLE MOSS (*Taxithelium kerianum*) A — whole plant; B — various leaves; C — capsule; D — leaf tip.

are complanate. The branch leaves are ovate with a long-tipped thread-like apex. They are concave, toothed, 1.3 × 0.4 mm. Leaf cells are wide, rhombic, and four to eight times as long as wide, each with a single papilla on the dorsal side. The base of the leaves have three to four pale yellow alar cells. The seta is smooth, red and 3.0–3.5 cm long. The capsule is horizontal, curved, and swollen on one side. The capsule lid is short and conical.

Localities: Common and widely distributed throughout Singapore and Malaysia on trees and logs. Also found in the Philippines, Nepal, Thailand, New Guinea, and Oceania.

*UNCOVERED NIPPLE MOSS

Taxithelium instratum

The Uncovered Nipple Moss forms dense, dull, yellowish green mats on the bark of trees and decaying wood. The main stem is creeping, with more or less pinnate side branches. The branch leaves are ovate with a contracted base. They are 1 × 0.4 mm, toothed, and short-tipped. Leaf cells are narrow, rhombic, thin-walled, about ten to twenty times as long as wide, and each with a row of papillae. In some cases the papillae are divided at their apex so that superficially there may appear to be two or three rows. The base of the leaf bears two distinct, inflated yellow alar cells with a few square cells above, forming a distinct group. The seta is 1.5 cm long and bears a horizontal capsule with a conical long-pointed lid.

Localities: Botanic Gardens, Jurong Road, Singapore; Gunong Tampin, Jerantut, Jor, Upper Perak; Kuala Rek, Kelantan. Also found in Thailand, Indonesia, New Guinea, and the Philippines.

*GOTTSCH'S NIPPLE MOSS

Taxithelium gottscheanum

The Gottsch's Nipple Moss is a robust, pale yellow-green moss forming lax, glossy mats. The stems are creeping, irregular pinnate with flat, blunt branches which are about 1 cm long. The leaves are broadly ovate with a constricted base, deeply concave and acute, 1.0 × 0.7 mm, and somewhat toothed above. The leaf cells are twelve times as long as wide with minute rows of papillae. There is a group of two to four small, oval, inflated alar cells. The seta is 1–2 mm long, and is very slender and smooth. The capsule is more or less erect with a conical lid.

Localities: Kranji, Bukit Timah, Singapore; East Malaysia. Also found in the Philippines.

***KERI'S NIPPLE MOSS**

Taxithelium kerianum (Fig. 90)

The Keri's Nipple Moss forms golden yellow, feathery tufts. The main stem is very long, and bears dense pinnate branches which are more or less complanate. The leaves are ovate-lanceolate, spreading, long acuminate, concave, entire, and about 1.0 × 0.3 mm. The leaf cells are narrow, linear, twenty times as long as wide, and with papillae in series over the lumen. The alar cells are rather small and few, but enlarged at the extreme leaf angles. The seta is 5 mm long with a minute, almost erect capsule. The lid is short-pointed.

Localities: Singapore, Perhentian Tinggi, Klang, Tapah, Penang, Kelantan. Also found in the Philippines, Australia, Java, and New Guinea.

***EQUAL BRANCH NIPPLE MOSS**

Taxithelium isocladum

The Equal Branch Nipple Moss is very similar to Keri's Nipple Moss but has a more robust habit, and bears very regular short pinnate branches. The alar cells are very small. The seta is 1 cm long. The capsule lid is longer than in Keri's Nipple Moss, being as long as

Localities: Frequent and recorded from all the states of West Malaysia and from Singapore.

***HAIRY NIPPLE MOSS**

Taxithelium capillipes

The Hairy Nipple Moss is a green shiny, creeping moss with the main stem with about 5 cm long, bearing more or less regular pinnate branches. The branches may be further divided in an irregular manner, and are 0.5—1.0 cm long. These branches are always flattened. The leaves are spreading, concave, about 1.5 × 1.0 mm, and have a broad oval base and a short tip. The leaf cells are long and smooth near the base, but with papillae near the tip. There is a group of square, bladder-like cells in each leaf corner. The length of the seta is 2 cm or less, red, smooth, and thin. The capsule is brown, ovoid, upright or hanging. The cap is conical and short-tipped.

Localities: Islands off Singapore, Bukit Timah; Port Swettenham; Fraser's Hill, Penang.

SIGNAL LEAF FAMILY

(Sematophyllaceae)

The Signal Leaf Family is a large family of creeping mosses with irregularly branched stems. The leaves are usually ovate and with a long drawn-out tip. The cells are linear, usually much longer than wide. They may be smooth or papillose. The leaf angles often have enlarged, coloured alar cells. The long seta may be smooth or papillose. The capsule is small, inclined, or horizontal. The double peristome is well developed.

REDDISH TERMINAL PORE MOSS

Acroporium rufùm (Fig. 91)

The Reddish Terminal Pore Moss is a small slender creeping moss. The stem bears irregular branches which may be hooked at their tips. The leaves are roughly in six rows. They are ovate or lanceolate in shape and have a wide pointed tip.

The seta is slender, 2 cm long, papillose. The capsule is narrow cylindrical with a definite annulus. The outer peristome teeth are much shorter than the inner ones. The capsule is slightly inclined.

Localities: Recorded from Malacca, Johore, Pahang, Perak, Penang, and Kedah, mainly over 1,000 m. It is highly variable but can often be distinguished by its papillose seta.

The **Rough-Leaved Terminal Pore Moss** (*Arcroporium asperifolium*) has been recorded from Seletar, Singapore. No specimen has been located.

*TRIPLE LARGE MOSS

Trismegistia lancifolia

The Triple Large Moss is a large wiry creeping moss, forming deep yellowish green to brownish tufts. The primary stem is very long and bears upright, copiously branched secondary branches, which are slightly complanate. Branch leaves are ovate lanceolate, concave and toothed above. There is a distinct border and a group of golden brown inflated alar cells at each leaf corner. The stem leaves are large and less toothed. The seta is 6 cm long. The capsule is borne horizontally. The long operculum has a slender beak.

Localities: Common throughout the area on trees and logs in damp forest. Also found in Indonesia, the Philippines, and Nepal.

*RIGID TRIPLE LARGE MOSS

Trismegistia rigida (Fig. 92)

The Rigid Triple Large Moss is a yellowish green moss, forming dense lax tufts. The main stem is creeping but bears rigid erect branches which are thickly pinnate above. The branch leaves are erect-spreading, ovate-lanceolate, about 2.5 cm long, and coarsely toothed, with an indistinct border and a long twisted tip. The leaf cells are elongated and about six times as long as wide. Groups of golden-brown, inflated alar cells are found at the leaf corners. The seta is red, about 0.8 cm long, and bears a horizontal capsule with a short conical lid.

Localities: Common and widely distributed throughout West Malaysia and Singapore. Also found in Thailand, Indonesia, and the Philippines.

*SMALL FRUIT MOSS

Meiothecium mircocarpum (Fig. 93)

The Small Fruit Moss forms flat, yellowish green, glossy mats with irregular pinnate, creeping stems. The leaves are oblong ovate, concave, plicate, 1.5 mm long, and have a short tip. The margins are entire. The leaf cells are oval-rhombic, five to six times as long as wide, minutely papillose at the apical angles of the cells walls, that is, not papillose over the lumen. The alar cells are large, hyaline, and coloured. The seta is 3—4 mm long, with a tiny inclined capsule 1.5 mm long. The lid is conical. This is probably the same as **Jagor's Small Fruit Moss** (*Meiothecium jagorii*) recorded from Singapore and Malacca.

Localities: Mostly on trees. Widespread in Singapore; recorded from Batu Caves and Penang. Also found in Nepal, Sri Lanka, the Philippines, New Guinea, and New Caledonia.

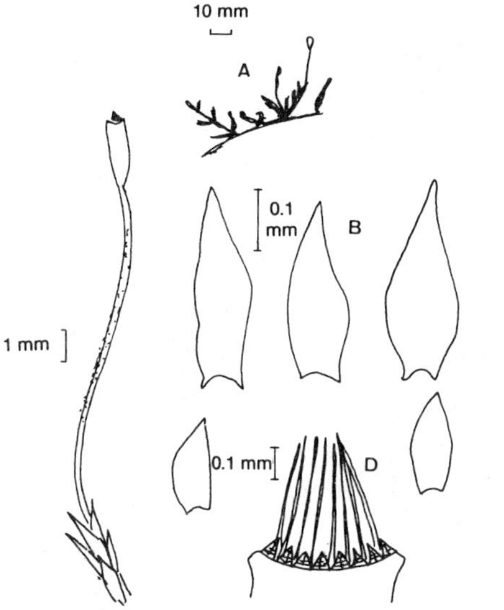

Fig. 91. REDDISH TERMINAL PORE MOSS (*Acroporium rufum*) A — whole plant; B — various leaves; C — capsule; D — peristome.

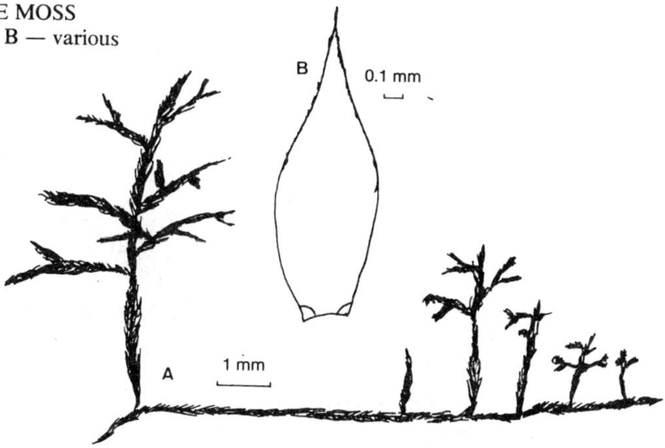

Fig. 92. RIGID TRIPLE LARGE MOSS (*Trismegistia rigida*) A — whole plant; B — leaf.

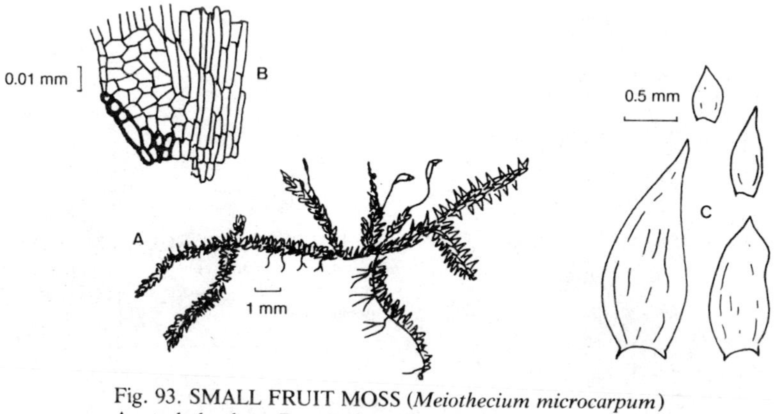

Fig. 93. SMALL FRUIT MOSS (*Meiothecium microcarpum*)
A — whole plant; B — leaf corner;
C — various leaves.

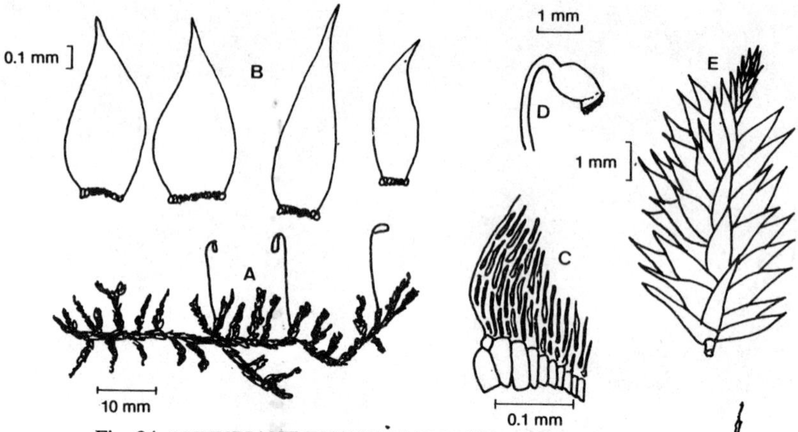

Fig. 94. LUXURIANT RHAPHĖ MOSS (*Rhaphidostichum luxurians*) A — whole plant; B — leaves; C — leaf corner; D—capsule; E—enlarged branch.

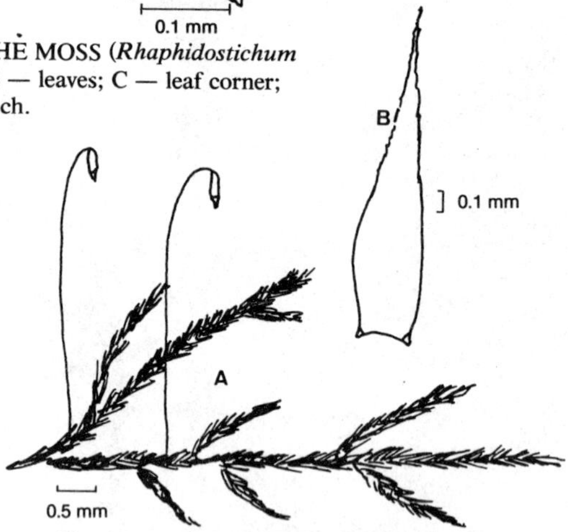

Fig. 95. BOSCH'S COLUMN MOSS (*Trichosteleum boschii*) A—whole plant; B—leaf.

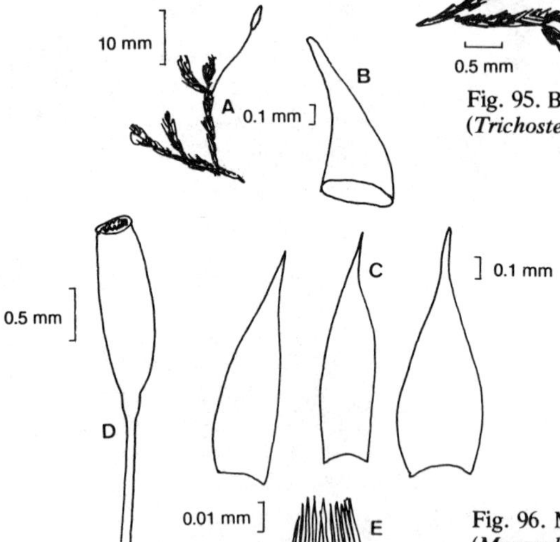

Fig. 96. MUELLER'S LARGE MEMBRANE MOSS (*Macrophymenium muelleri*) A — whole plant; B — operculum; C — leaves; D — capsule; E — peristome.

*WARBURG'S NARROW FRUIT MOSS

Warburgiella leptocarpa

The Warburg's Narrow Fruit Moss is a dull yellow to dirty green moss. The main stem is creeping, with numerous branches arranged roughly in a pinnate manner. The branches are single and 1.5 cm long. Branch leaves are concave, 1.5 × 0.3 mm, spoon-shaped, with a narrowed base and a rounded tip bearing prominent teeth. Leaf cells are linear and about three times as long as wide with papillae at the cells corners, not over the lumen. Leaf corners bear about two thick-walled alar cells.

Localities: On tree bark, Singapore. Also found in Indonesia and Sri Lanka.

*ROTTEN WOOD SIGNAL MOSS

Sematophyllum saproxylophilum

The Rotten Wood Signal Moss is a delicate, slender, creeping moss forming dense, pale green tufts on tree bark or logs. The main stem is rather short and irregularly branched. The leaves are rigid, erect-spreading, narrow lanceolate, and about 2 × 0.4 mm. The margin is minutely toothed near the apex. The leaf cells are linear and about fifteen times as long as wide, and not much thickened. The leaf corners have three or four alar cells plus some smaller cells above. The seta is smooth, 8 mm long, curved at the tip, and bears a small inclined capsule

Localities: Seletar, Bukit Timah, Singapore; Gunong Belumut, Kuala Trengganu; Gunong Sitong, Kelantan. Also found in Java and the Philippines

*SHORT-BRANCHED SIGNAL MOSS

Sematophyllum microcladiellum

The Short-Branched Signal Moss is a creeping moss with the main stem bearing numerous side branches. The branch leaves are oval to broad lanceolate, 1.4 × 0.5 mm, concave, short-tipped, and with an entire margin. The leaf cells are thin-walled, about eight times as long as wide, rhombic, and transparent, without papillae. The alar cells are large and prominent, golden in colour, thin-walled, rectangular to elliptic. The seta is smooth, 1.5 cm long, and bears a long horizontal capsule.

Localities: Kranji, Woodlands, Jurong, Singapore. Also found in Indonesia and New Guinea.

*LUXURIANT RHAPHE MOSS

Rhaphidostichum luxurians (Fig. 94)

The Luxuriant Rhaphe Moss is a beautiful luxuriant moss forming thick, golden green tufts. The main stem is dark in colour, long creeping, and up to 0.4 mm in section. Branches are 1–2 cm long, rarely sparingly branched themselves, and somewhat flattened. Branch leaves are concave, oval, but suddenly narrowed and drawn out at the tip. The leaf cells are thickened, rhombic to elliptical, and about ten times as long as wide. Alar cells are prominent, forming round groups of about five cells. Cells of the insertion across the base of the leaf are golden yellow in colour. The seta is rough and 1–1.5 cm high, bearing a horizontal to pendulous capsule.

Localities: Singapore; Mt. Ophir; Penang. Also found in Indonesia.

*BRUCH'S RHAPHE MOSS

Rhaphidostichum bruchii

The Bruch's Rhaphe Moss is very similar to the previous species, but the leaves tend to be shorter tipped. The leaf cells are papillose, and the seta is always smooth.

Localities: Singapore; Gunong Panti, Petaling, Kuala Tahan, Kuala Lipis, Tapah, Gunong Keledang, Penang. Also found in Indonesia.

*BOSCH'S COLUMN MOSS

Trichosteleum boschii (Fig. 95)

The Bosch's Column Moss forms loose, golden green, somewhat shiny tufts. The stems are long creeping, dark brown, and irregularly pinnate. The branches are ascending, with complanate leaves. Branch leaves are ovate-lanceolate, slightly concave, and gradually or abruptly long acuminate, 2 × 0.5 mm. The margins are often toothed near the apex, and reflexed. Leaf cells are linear-rhombic, fifiteen times as long as wide, with a single papilla over the lumen on the dorsal side. The seta is about 1 cm long, curved, and papillose at the top, smooth below, and bearing a small, horizontal, or pendulous capsule.

Localities: This species is common and widely distributed throughout Singapore and West Malaysia. It is also found in Thailand, Indonesia, and Fiji. **Singapore Column Moss** (*Trichosteleum singapurense*) appears to be identical with this species, except the seta is less rough and the leaf cells less papillose. **One-Row Column Moss** (*Trichosteleum monostictum*) which is also probably not a distinct species has longer and more flexuose leaves.

◆ ◆ ◆

RUINED TOOTH MOSS FAMILY
(Rhegmatodontaceae)

The Ruined Tooth Moss Family is a small family of medium-sized, yellowish green mosses. The leaves are ovate lanceolate, with a drawn-out tip. The creeping stem is irregularly branched. The capsule is exerted, erect, and cylindrical. The peristome is double, with the inner peristome well developed and over-topping the outer peristome.

MUELLER'S LARGE MEMBRANE MOSS
Macrohymenium muelleri (Fig. 96)

The Mueller's Large Membrane Moss is a small to medium sized moss with a sparsely branching prostrate stem giving rise to upright secondary stems which may also become detached and form rhizoids. The leaves are lanceolate, with a long drawn-out tip. They are very crowded near the base of the seta. The seta is about 1 cm long, and somewhat wavy with a slight swelling below the capsule. The outer peristome is very short. Behind this peristome the long projecting inner peristome segments can be seen. The operculum is asymmetrical with a wide tip.

Localities: Gunong Tahan. Also found in Indonesia.

GLOSSARY

Acrocarpous	having capsule terminal on main stem
acute	pointed without sudden narrowing.
alar cells	specialized cells at basal angles of leaf
amphigastria	underleaves of leafy liverworts
annulus	ring of cells around the mouth of a capsule
antheridium	male reproductive organ
apical	at the end of a stem or branch
adpressed	held close to stem and not spread out
archegonium	female reproductive organ
aristate	with a fine, hair-like point
ascending	pointing upwards; becoming erect
auricle	a small ear-like lobe at basal angles of a leaf
axis	an imaginary line down the centre of the stem
basal angles	the corner of the leaf next to the stem (The leaf is treated as if it were an imaginary isosceles triangle.)
basal leaf cells	cells of the lamina near to the stem
basal membrane	a continuous tissue uniting inner peristome segments
beak	drawn-out point on lid, resembling a bird's beak
bipinnate	twice pinnate
border	specialized marginal region, differing in kind of cells
bryoid	with a double peristome and high basal membrane with cilia alternating the inner teeth
calyptra	protective covering found on the top of a young capsule
capsule	reproductive organ containing spores; "fruit"
cell	microscopic, walled unit of structure
chlorocyst	a cell containing green chloroplasts
chloroplast	green pigment-containing body within the cell
cilia	hair-like threads
coma	a tuft of leaves at the tip of the stem; *comal* (adj.)
complanate	more or less flattened in one plane

crenate	with coarse, rounded teeth
crenulate	with fine rounded teeth.
cygneous	suddenly curved like a swan's neck
cylindric	shaped like a cylinder
dentate	with sharp, pointed teeth at 90° to the margin
dicranoid	with a single row of bifid peristome teeth
dimorphic	of two forms
dorsal	at the back; the face of a leaf remote from the stem
double peristome	peristome teeth in two rows
emergent	held above the perichaetial leaves on a long seta
entire	without teeth or divisions
epiphragm	a specialized membrane across the mouth in Polytrichaceae
epiphyte	growing on another plant but not parasitic
erect	held upwards or upright; at a small angle to the stem
erect-spreading	with basal of leaf erect and apex spreading away from stem
excurrent	with mid-rib running beyond leaf lamina
exerted	elevated on a long seta, high above perichaetial leaves
falcate	curved like a sickle
fascicles	in bunches or bundles
fibrillose	with fine fibres in the cells, e.g., *Sphagnum*
filiform	thread-like
flexuose	readily bent
"fruit"	popular name for capsule
gemma	bud capable of producing plants directly
glaucous	bluish greyish green
gregarious	with groups of plant growing together
homotropous	with leaves all growing in the same direction
horizontal	(of capsule) held at right angles to the seta
hyaline	transparent; not green
hyalocyst	dead colourless cell with pores and fibrillose thickenings

immersed	buried among the perichaetial leaves; not elevated on seta
incrassate	with thick walls
inclined	held at a small angle to the seta
inflated	swollen like a balloon
innovation	forming new branches below the apex
iridescent	shining to give rainbow hues
isodiametric	with equal dimensions in all directions
lamellae	with short upright columns of cells on surface
lamina	expanded blade of leaf other than mid-rib
lanceolate	with rounded wide base and long pointed tip, like a lance
leucocyst	dead colourless cell without spiral thickenings
lingulate	tongue-shaped
linear	very narrow with near-parallel sides
lumen	part of cell not immediately above cell wall
mid-rib	nerve; central differentiated portion more than one cell thick
mitriform	shaped like a bishop's mitre; cap-shaped with a fringe
mouth	hole at the top of the capsule when the lid is removed
mucronate	with a short, stout, abrupt point
neck	part of capsule where it joins seta
nodulose	with knots or swellings
oblique insertion	joined at an oblique angle to the axis
operculum	lid of capsule
ovate, ovoid	egg-shaped, with widest part at base
papilla, papillae	minute, acute, or rounded protruberances (adj.) *papillose, papillate*
paraphyllia	leaf-like, freely branched structures borne among ordinary leaves in some mosses, e.g., *Pelekium*
parenchymatous	composed of isodiametric or rectangular cells
pendent	hanging
pendulous	loosely hanging
perichaetial leaves	special leaves round the sex organs and base of seta

perichaetium	perichaetial leaves
peristome	a fringe of teeth round the mouth of the capsule of most mosses
persistent	not falling off
pleurocarpous	creeping and bearing fruit on very short lateral branches
plicate	with longitudinal folds
pinnate	with side branches arranged like a feather
primary stem	main stem from which others may arise
procumbent	trailing or prostrate
prosenchymatous	composed of elongated cells with oblique cross walls
protonema, protonemata	green branched thread formed from spores
pseudonerve	false nerve down centre of true nerve in *Leucophanes*
quadratic	approximately square
radial	symmetrical or formed on the radii of an imaginary circle
rhombic, rhomboid	diamond-shaped
rhizoids	thread-like outgrowths for support or attachment
scabrous	rough
secondary stem	branch of main (primary) stem
seta	stalk of the capsule
sheath	specially differentiated basal part of leaf next to stem
shoulder	curve formed by sudden narrowing of leaf from broad base
simple stem	unbranched stem
single peristome	with peristome teeth in one row
sinuose	wavy
solitary	growing as single plants
spermatozoid	tiny, motile, male reproductive cell
spinulose	with small spines
spore	a tiny unit of non-sexual reproduction
spreading	leaves turned at an obtuse angel to the stem
stomates	holes in wall of capsule, protected by guard cells
stomium	mouth of capsule
striate	having faint furrows (*striae*)
subulate	shaped like an awl

teniolae	special thickened cells found in a row a short way from margin
terminal	at the end of the stalk or seta
thallus	an undifferentiated plant without stem and leaves
translucent	transmitting light through the leaf
tripinnate	three times pinnate
truncate	with tip suddenly cut off
undulate	wavy
vaginula	a minute sheath at the base of the seta
ventral	at the front, i.e., the face of the leaf nearest to the stem

APPENDIX: HINTS ON THE MEANINGS OF NAMES

Hints on the Meanings of Family Names

(G) = Greek; (L) = Latin

Bryaceae	*bryon* (G) old Greek plant name
Calymperaceae	*kalymma* (G) hood, cover; *pera* (G) pouch, wallet
Dicranaceae	*dicranon* (G) two-pronged forks
Fissidentaceae	*fissus* (L) split; *dens* (L) tooth
Hookeriaceae	after Sir W.J. Hooker, nineteenth-century botanist and author
Hypnaceae	*hypnon* (G) an old Greek plant name
Leucobryaceae	*leukos* (G) white; *bryon* (G) old Greek plant name
Neckeraceae	after N.J. Necker, eighteenth-century botanist
Orthotrichaceae	*ortho* (G) upright; *tricho* (G) hair
Polytrichaceae	*poly* (G) many; *tricho* (G) hair
Pottiaceae	after Professor Pott, botanist in Brunswick
Rhegmatodontaceae	*rhegma* (G) a ruin; *odous* (G) tooth
Rhizogoniaceae	*rhiza* (G) root; *gone* (G) offspring
Sematophyllaceae	*semato* (G) mark, sign, signal; *phyllon* (G) a leaf
Sphagnaceae	*sphagnos* (G) an old plant name
Syrrhopodontaceae	*syrrhopos* (G) horizontal; *odous* (G) tooth
Thuidiaceae	*Thuja* (L) botanical name of a coniferous tree
Weberaceae	from Weber, a famous botanist

Hints on the Meanings of Species Names

Acanthorrhynchium papillatum	*acantho* (G) spiny, thorny; *rhyncho* (G) with snout or beak; *papillatum* (L) with papillae
Acroporium asperifolium	*akros* (G) tip, end; *porous* (G) pore, opening; *asper* (L) rough; *folium* (L) leaf
A. rufum	*rufum* (L) red
Aerobryopsis longissima	*aeros* (G) air; *bryon* (G) old plant name; *longissima* (L) very long
Arthrocormus schimperi	*arthro* (G) jointed; *kormos* (G) corm; *schimperi*—after W.P. Schimper
Barbula louisiadum	*barbula* (L) a little beard; *louisiadum* — from Louisiaden, a place name
Braunfelsia dicranoides	*Braunfelsia* — after a famous botanist; *dicranoides* — like *Dicranum,* a genus of mosses
B. enervis	*enervis* (L) without a nerve
Bryum coronatum	*bryon* (G) an old plant name; *corona* (L) a crown
Callicostella prabaktiana	*kalli* (G) beautiful; *costella* (L) a small nerve; *prabaktiana* — derived from a place name
Calymperes dozyanum	*kalymma* (G) hood, cover; *pera* (G) pouch, wallet; *dozyanum* — after F. Dozy, a botanist
C. hampei	after E. Hampe, a botanist
C. longifolium	*longi* (L) long; *folium* (L) leaf
C. nicobarense	from the Nicobar Islands
C. porrectum	*porrectum* (L) stretched out and forwards
C. recurvifolium	*recurvi* (L) curved backwards; *folium* (L) leaf
C. serratum	*serratum* (L) toothed
C. tahitense	from Tahiti
C. tenerum	*tenerum* (L) delicate
Calymperidium tjibodensis	*kalymma* (G) hood, cover; *pera* (G) pouch, wallet; *tjibodensis* from Tjibodas, a place in Indonesia.

Campylodontium flavescens	*kampylo* (G) bent; *odont* (G) tooth; *flavescens* (L) yellowish
Campylopodium euphorocladum	*kampylo* (G) bent; *podium* (G) foot; *euphoros* (G) true bearing; *klados* (G) branch or shoot
Campylopus comosus	*kampylo* (G) bent; *pous* (G) foot; *comosus* (L) bearing a tuft of leaves at the apex
C. richardii	after P.W. Richards, a modern botanist
C. serratus	*serratus* (L) toothed
C. thwaitsii	after G.H.K. Thwaites, a botanist
Chaetomitrium elongatum	*chaete* (G) hair; *mitrium* (G) cap; *elongatum* (L) elongated
C. leptopoma	*lepto* (G) slender; *poma* (L) fruit
C. orthorrhynchum	*ortho* (G) straight, upright; *rhynchos* (G) snout, beak
Chionoloma latifolium	*chion* (G) snow; *loma* (G) hem, border; *latifolium* (L) broad-leaved
Cladopodanthus pilifer	*klados* (G) branch, shoot; *podion* (G) foot; *anthos* (G) flower; *pilifer* (L) bearing hairs
Desmotheca apiculata	*desmos* (G) a band; *theke* (G) case, chest; *apiculata* (L) long pointed
Dicranella coarctata	*dikranella* (G) a small two-pronged fork; *coarctata* (L) compressed
Dicranodontium nitidum	*dikrano* (G) two-pronged fork; *odont* (G) tooth; *nitidum* (L) brilliant; shining
Dicranoloma assimile	*dikrano* (G) two-pronged fork; *loma* (G) border, hem; *assimile* (L) like, similar
Diphyscium mucronifolium	*di* (G) two; *physo* (G) bladder; *mucronifolium* (L) with a mucronate leaf (L)
Distichophyllum cuspidatum	*di* (G) two; *stichos* (G) row; *phyllon* (G) leaf; *cuspis* (L) a point
D. gracilicaule	*gracilis* (L) slender, thin; *caulis* (L)
D. schmidtii	after Schmidt
D. spathulatum	*spathulatum* (L) spoon-shaped
Ditrichum difficile	*di* (G) two; *tricho* (G) hair *difficile* (L) difficult
Ectropothecium buitenzorgii	*ek* (G) outside; *tropo* (G) folded; *thekion* (G) a little vessel *buitenzorgii* — from Buitenzorg, the old Dutch name for Bogor, Indonesia.

E. incubans	*incubans* (L) obliquely inserted
E. monumentorum	*monumentorum* (L) of a monument
E. moritzii	after Moritz
E. singapurense	from Singapore
Endotrichella elegans	*endo* (G) inside; *trichella* (G) a little hair; *elegans* (L) elegant
Exodictyon blumii	*exo-* (G) outwards; *diktyon* (G) a net; *blumii* — after Blume
E. sullivantii	after W.S. Sullivant
Fissidens ceylonensis	*fissus* (L) cleft; *dens* (L) tooth; *ceylonensis* — from Ceylon (Sri Lanka)
F. crassinervis	*crassus* (L) thick; *nervis* (L) nerved
F. mittenii	after W. Mitten
F. sylvaticus	*sylvaticus* (L) woodland
Floribundaria floribunda	*floribundus* (L) profusely flowering
Garckia phascoides	*Garckia* — after Garck; *phascon* (G) a plant name
Himantocladium plumula	*himanto* (G) strap-shaped; *klades* (G) stem; *plumula* (L) feather
Hydrogonium arcuatum	*hydro* (G) water; *gonium* (L) reproductive organ; *arcuatum* (L) curved like a bow
Hymenostomum edentulum	*hymen* (G) membrane; *stoma* (L) mouth; *edentatus* (L) toothless
Hyophila involuta	*hyo* (G) water, contraction of *hydro;* *philo* (G) loving; *involutus* (L) rolled inwards
H. javanica	from Java
Isopterygium albescens	*iso* (G) equal; *pterygo* (G) wings; *albescens* (L) white
I. bancanum	from Banka, an island
I. minutirameum	*minutus* (L) very small, minute; *rameus* (L) belonging to a branch
Leucobryum aduncum	*leukos* (G) white; *bryon* (G) an old plant name; *aduncum* (L) bent inwards
L. sanctum	*sanctum* (L) holy
L. javense	from Java
Leucoloma molle	*leukos* (G) white; *loma* (G) border; *molle* (G) soft, pliable
Leucophanes albescens	*leukos* (G) white; *phaneros* (G) conspicuous; *albescens* (L) white
L. candidum	*candidum* (L) pure
L. octoblepharioides	*octo* (L) eight; *blepharis* (G) eyelash

Macrohymenium muelleri	*makro* (G) large; *hymen* (G) membrane; *muelleri*—after C. Mueller
Macromitrium incurvifolium	*makro* (G) large; *mitrium* (L) cap; *incurvifolium* (L) with curved-in leaves
Meiothecium microcarpum	*meio* (G) small; *theke* (G) vessel; *micro* (L) small; *carpon* (L) fruit
M. jagorii	after Jagor
Meteorium miquelianum	*metior* (L) a measure; *miquelianum* — after Miquel
Microdus miquelianus	*mikro* (G) small; *odous* (G) tooth; *miquelianus* — after Miquel
Octoblepharum albidum	*octo* (L) eight; *blepharis* (G) eyelash; *albidum* (L) white
Pelekium bifarium	*pelekus* (G) hatchet, axe; *bifarium* (L) in two rows
P. velatum	*velatum* (L) covered
Pinnatella mucronata	*pinnatella* (L) like a little feather; *mucronata* (L) mucronate
Plagiothecium neckeroides	*plagio* (G) oblique; *thekion* (G) a little vessel; *neckeroides* — after N.J. Necker
Pogonatum macrophyllum	*pogon* (G) a beard; *makro* (G) large; *phyllum* (G) leaf
P. junghuhnianum	after F.E. Junghuhn
Racelopus pilifer	*rhakos* (G) a ruin; *opous* (G) tooth; *pilifer* (L) bearing hairs
Rhaphidostichum bruchii	*rhaphido* (G) rhaphe; *stichum* (G) row; *bruchii* — after Bruch
R. luxurians	*luxurians* (L) luxuriant
Rhizogonium latifolium	*rhiza* (G) root; *gonë* (G) offspring; *latifolium* (L) broad leaf
R. spiniforme	*spini* (L) spine; *forme* (L) form
Schistomitrium apiculatum	*schistos* (G) cleft; *mitrium* (G) cap; *apiculatum* (L) long pointed
Sematophyllum microcladiellum	*semato* (G) signal; *phyllum* (G) leaf; *mikro* (G) small; *kladiellum* (G) short shoot
Sematophyllum saproxylophilum	*sapro* (G) rotten; *xylo* (G) wood; *philum* (G) loving
Semibarbula orientalis	*semi* (L) half; *barbula* (L) a little beard; *orientalis* (L) Eastern
Sphagnum flaccidifolium	*sphagnos* (G) old plant name; *flaccidus* (L) flaccid; *folium* (L) leaved
S. junghuhnianum	after F.E. Junghuhn

Symphysodon neckeroides	*symphysis* (G) joining; *odous* (G) tooth; *neckeroides* — after N.J. Necker
Syrrhopodon albo-vaginatus	*syrrhopos* (G) horizontal; *odous* (G); *albo* (L) white; *vaginatus* (L) sheathing
S. ciliatus	*ciliatus* (L) with fine hairs
S. confertus	*confertus* (L) compact
S. croceus	*croceus* (L) saffron yellow
S. gardneri	after G. Gardner
S. muelleri	after C. Mueller
S. revolutus	*revolutus* (L) rolled backwards
S. ridleyi	after Ridley, a Singapore botanist
S. rufescens	*rufescens* (L) reddish
S. spiculosus	*spiculosus* (L) covered with fine points
Taxithelium capillipes	*taxis* (G) order; *thele* (G) nipple, teat; *capillaceus* (L) thread-like
T. gottscheanum	after Gottsch
T. instratum	*instratum* (L) uncovered
T. isocladum	*iso* (G) equal; *klados* (G) branch
T. kerianum	after Keri
Thyridium cardotii	*thyris* (G) door; *cardotii* — after Cardot
T. fasciculatum	*fasciculatum* (L) in bundles
T. flavum	*flavum* (L) yellow
T. manii	after Man
T. repens	*repens* (L) creeping
T. undulatum	*undulatum* (L) wavy
Trichosteleum boschii	*trichos* (G) hair; *steleum* (L) column; *boschii* — after Bosch
T. monostichum	*mono* (L) one; *stichum* (L) row
T. singapurense	from Singapore
Trichostomum sarawakensis	*trichos* (G) hair; *stomum* (G) mouth *sarawakensis* — from Sarawak
Trismegistia lancifolia	*tris* (G) three times; *mega* (G) large; *lancifolia* (L) with lance-shaped leaves
T. rigida	*rigida* (L) rigid
Vesicularia dubyana	*vesicularia* (L) bladder; *dubyana* (L) dubious
V. kurzii	after Kurz
V. miquelii	after Miquel
V. reticulata	*reticulata* (L) net
Warburgiella leptocarpa	*Warburgiella* — after Warburg; *leptos* (G) narrow; *carpa* (L) fruit
Wilsoniella pellucida	*Wilsoniella* — after W. Wilson; *pellucida* (L) transparent but not hyaline

REFERENCES

Bartram, E.B. (1939) Mosses of the Philippines. *Phil. J. Sc.* 68: 1−437.

Burkhill, I.H. (1966) A Dictionary of the Economic Products of the Malay Peninsula, vol. 1, p. 420. 2nd ed. Ministry of Agriculture and Cooperatives, Kuala Lumpur.

Chopra, R.S. (1975) *Taxonomy of Indian Mosses* (*An Introduction*), Publication and Information Directorate (CSIR), New Delhi.

Dixon, H.N. (1924) New Species of Mosses from the Malay Peninsula. *Bull. Torrey Bot. Club* 51: 225—59.

Dixon, H.N. (1926) A List of the Mosses of the Malay Peninsula. *Gard. Bull., Str. Settl.* 4: 1−46.

Dozy, F., and Molkenboer, J.H. (1854) Musci Frondosi Inediti Archipelagi Indici.

Dozy, F. and Molkenboer, J.H. (1855−70) Bryologia Javanica, vols. 1−2.

Fleischer, Max (1904−1922) die Musci der Flora von Buitenzorg, vols. 1−4, E.J. Brill, Leiden.

Johnson, Anne (1959) The Genus Sphagnum in Malaysia. *Gard. Bull., Str. Settl.* 17: 312−24.

Johnson, Anne (1962) On the Widespread Occurrence of a Sand-Binding Moss on the East Coast of Malaysia. *Mal. Nat. J.* 16: 111−12.

Johnson, Anne (1964) An Account of the Malaysian Leucobryaceae. *Gard. Bull., Str. Settl.* 20: 315−60.

Johnson, Anne (1969) Classification of Mosses and Liverworts of Singapore Island, *Nanyang Univ. J.* 3: 430—37.

Johnson, Anne (1973) The Fissidentaceae (Split-Teeth Mosses) of Singapore, with Notes on Their Malaysian Distribution. *Nanyang Univ. J.* 7: 64—77

Mitten W. (1859) Musci Indiae Orientalis: An Enumeration of Mosses of the East Indies. *J. Proc. Linn. Soc.* (*Bot.*) 1: 1−96.

Tixier, P. (1971) A Contribution to the Bryological Knowledge of Fraser's Hill Station (Malaysia). *Gard. Bull., Sing.* 25: 335−53.

Van der Wijk, R. (1958) Precursory Studies on Malaysian Mosses. II: A Preliminary Key to the Moss Genera. *Blumea* 9: 143−86.

Van de Wijk, R. (ed.) (1959−69) Index Muscorum, vol. 1−5. Utrecht.

Van der Wijk, R., and Chopra, R.S. (1966) A Preliminary Key to the Genera of Indian Mosses. *Research Bull., Punjab Univ.* 17: 149−91.

INDEX

Acanthorrhynchium papillatum, 12, 103, **104**, 119
Acroporium asperifolium, 12, 107, 119
A. rufum, 12, 107, **109**, 119
Aerobryopsis longissima, 11, **85**, 87, 119
Arthrocormus schimperi, 8, 42, **44**, 119
Barbula louisiadum, 10, 74, **75**, 119
Braunfelsia dicranoides, 8, 30, **32**, 119
B. enervis, 8, **32**, 33, 119
Bryaceae, 10, 77, 118
Bryum coronatum, 10, **76**, 77, 119
Callicostella prabaktiana, 11, 94, **95**, 119
Calymperaceae, 9, 49, 118
Calymperes delessertii, 7
C. dozyanum, 9, **48**, 50, 119
C. hampei, 9, 52, **53**, 119
C. longifolium, 9, **48**, 51, 119
C. nicobarense, 9, 52, **54**, 119
C. porrectum, 9, 51, **53**, 119
C. recurvifolium, 9, 51, **53**, 119
C. serratum, 9, **48**, 50, 119
C. tahitense, 9, **52**, 53, 119
C. tenerum, 9, **54**, 55, 119
Calymperidium tjibodensis, 9, **54**, 55, 119
Campylodontium flavescens, 11, 89, **91**, 120
Campylopodium euphorocladium, 8, 30, **31**, 120
Campylopus comosus, 8, 34, **35**, 120
C. richardii, 8, **36**, 38, 120
C. serratus, 8, **35**, 37, 120
C. thwaitsii, 8, **36**, 37, 120
Chaetomitrium elongatum, 11, 94, 95, 120
C. leptopoma, 11, **92**, 93, 120
C. orthorrhynchum, 11, **92**, 93, 120
Chionoloma latifolium, 10, **69**, 71, 120
Cladopodanthus pilifer, 8, **44**, 46, 120
Desmotheca apiculata, 11, **81**, 82, 120
Dicranaceae, 8, 24, 118
Dicranella coarctata, 8, 29, **31**, 120
Dicranodontium nitidum, 8, **36**, 38, 120
Dicranoloma assimile, 8, **32**, 33, 120
Diphyscium mucronifolium, 10, 79, 120
Distichophyllum cuspidatum, 11, 89, **91**, 120
D. gracilicaule, 11, 90, 120
D. schmidtii, 11, 90, **92**, 120
D. spathulatum, 11, 90, **91**, 120
Ditrichum difficile, 8, **26**, 28, 120
Ectropothecium buitenzorgii, 11, 97, **99**, 120
E. incubans, 12, 98, **100**, 121
E. monumentorum, 12, **100**, 101, 121
E. moritzii, 11, 98, **99**, 121
E. singapurensis, 11, 98, 121
Endotrichella elegans, 11, 83, **85**, 121
Exodictyon blumii, 9, 46, **47**, 121
E. sullivantii, 9, 46, **47**, 121
Fissidens ceylonensis, 10, **66**, 67, 121
F. crassinervis, 2, 10, **66**, 68, 121
F. mittenii, 10, 68, **69**, 121
F. sylvaticus, 10, **66**, 67, 121
Fissidentaceae, 10, 64, 118
Floribundaria floribunda, 11, **86**, 88, 121
Garckia phascoides, 8, **26**, 27, 121
Himantocladium plumula, 11, 84, **85**, 121
Hookeriaceae, 11, 88, 118
Hydrogonium arctuatum, 10, 74, **75**, 121
Hymenostomum edentulum, 10, **69**, 72, 121
Hyophila involuta, 10, 73, 121
H. javanica, 10, **70**, 72, 121
Hypnaceae, 11, 97, 118
Isopterygium albescens, 12, **100**, 101, 121
I. bancanum, 12, 102, 121
I. minutirameum, 12, 101, 121
Leucobryaceae, 8, 39, 118
Leucobryum aduncum, 8, 13, 40, **43**, 121
L. sanctum, 8, 13, 39, **43**, 121
L. javense, 8, 13, 40, 121
Leucoloma molle, 8, 34, **35**, 121
Leucophanes albescens, 8, 42, **44**, 121
L. candidum, 8, 41, 121
L. octoblepharoides, 7, 8, 41, **43**, 122
Macrohymenium muelleri, 12, **110**, 113, 122
Macromitrium incurvifolium, 11, 83, 122
Meiothecium microcarpum, 12, 108, **109**, 122
M. jagori, 12, 108, 122
Meteorium miquelianum, 11, **86**, 87, 122
Microdus miquelianus, 8, 28, **31**, 122
Neckeraceae, 11, 83, 118
Octoblepharum albidum, 9, **47**, 49, 122
Orthotrichaceae, 11, 82, 118
Pelekium bifarium, 11, 96, 122
P. velatum, 11, **95**, 96, 122
Pinnatella mucronata, 11, 86, **88**, 122
Plagiothecium neckeroides, 11, 97, **99**, 122
Pogonatum junghuhnianum, 10, 80, 122
P. macrophyllum, 10, 79, **81**, 122
Polytrichaceae, 10, 79, 118
Pottiaceae, 10, 71, 118
Racelopus pilifer, 10, 80, **81**, 122
Rhaphidostichum bruchii, 12, 112, 122
R. luxurians, 12, **110**, 112, 122
Rhegmatodontaceae, 12, 113, 118
Rhizogoniaceae, 10, 77, 118
Rhizogonium latifolium, 10, 78, 122
R. spiniforme, 10, **76**, 78, 122
Schistomitrium apiculatum, 8, **44**, 45, 122
Sematophyllaceae, 12, 107, 118

Sematophyllum microcladiellum, 12, 111, 122
S. sapoxylophilum, 12, 111, 122
Semibarbula orientalis, 10, **70**, 73, 122
Sphagnaceae, *Sphagnum*, 2, 8, 23, 118
Sphagnum flaccidifolium, 8, 23, 24, **25**, 123
S. junghuhnianum, 8, 24, **25**, 123
Symphysodon neckeroides, 11, 84, **85**, 123
Syrrhopodon albo-vaginatus, 9, 57, 123
S. cilatus, 9, 58, **59**, 123
S. confertus, 9, 62, 123
S. croceus, 9, 60, **61**, 123
S. gardneri, 9, 56, **59**, 123
S. muelleri, 9, **60**, 61, 123
S. revolutus, 9, 56, **59**, 123
S. ridleyi, 9, 61, 123
S. rufescens, 9, 57, **59**, 123
S. spiculosus, 9, 58, **60**, 123
Syrrhopodontaceae, 9, 55
Taxithelium capillipes, 12, 106, 123
T. gottscheanum, 12, 105, 123
T. instratum, 12, 105, 123
T. isocladum, 12, 106, 123
T. kerianum, 12, 104, **106**, 123
Thuidiaceae, 11, 94, 118
Thyridium cardotii, 9, 63, 123
T. fasciculatum, 9, 62, **65**, 123
T. flavum, 9, 62, **65**, 123
T. manii, 9, 64, 123
T. repens, 9, 63, 123
T. undulatum, 9, 64, **65**, 123
Trichosteleum boschii, 12, **110**, 112, 123
T. monostichum, 12, 113, 123
T. singapurense, 12, 113, 123
Trichostomum sarawakensis, 10, 74, **75**, 123
Trismegistia lancifolia, 12, 107, 123
T. rigida, 12, 108, **109**, 123
Vesicularia dubyana, 12, 103, **104**, 124
V. kurzii, 12, 103, 124
V. miquelii, 12, 102, 124
V. reticulata, 12, 102, 124
Warburgiella leptocarpa, 12, 111, 124
Weberaceae, 10, 78, 118
Wilsoniella pellucida, 8, **26**, 27, 124